T0342164

Defected Ground Structure (DGS) Based Antennas

Defected Ground Structure (DGS) Based Antennas

Design Physics, Engineering, and Applications

Debatosh Guha, Chandrakanta Kumar, and Sujoy Biswas

IEEE PRESS
WILEY

Published by John Wiley & Sons, Inc., Hoboken, New Jersey.
Published simultaneously in Canada.

For general information on our other products and services or for technical support, please contact our Customer Care Department within the United States at (800) 762-2974, outside the United States at (317) 572-3993 or fax (317) 572-4002.

Wiley also publishes its books in a variety of electronic formats. Some content that appears in print may not be available in electronic formats. For more information about Wiley products, visit our web site at www.wiley.com.

Library of Congress Cataloging-in-Publication Data Applied for:

ISBN: 9781119896180 (hardback)

Cover Design: Wiley
Cover Image: © zf L/Getty Images

Set in 9.5/12.5pt STIXTwoText by Straive, Chennai, India

Contents

Author Biographies

Debatosh Guha is a Professor in Radio Physics and Electronics, University of Calcutta and Abdul Kalam Technology Innovation National Fellow. He is a Fellow of the IEEE and also of the national academies of his country such as the Indian National Science Academy (INSA), Indian Academy of Sciences (IASc), The National Academy of Sciences, India (NASI), and Indian National Academy of Engineering (INAE). He has served *IEEE Transactions on Antennas and Propagation* and *IEEE Antennas and Wireless Propagation Letters* as an associate editor for two consecutive terms. He has served on the IEEE AP-S Fields Award Committee (2017–2019) and is serving the IEEE AP-S MGA Committee from 2021 as a member and is chairing the IEEE Technical Committee on Antenna Measurements since 2022. He has represented India in the URSI Commission B since 2015. He co-edited a book titled *Microstrip and Printed Antennas: New Trends, Techniques and Applications* in 2010, published by Wiley. He is a Distinguished Lecturer of the IEEE Antennas and Propagation Society.

Chandrakanta Kumar received his M. Tech and PhD in Radio Physics and Electronics from the University of Calcutta, and completed the Space Studies Program, from International Space University, Strasbourg, France. He is an engineer with U R Rao Satellite Centre and primarily responsible for designing antennas for satellites and ground stations. His designs flew onboard Chandrayaan-1 and 2, Moon Impact Probe, Mars Orbiter Mission, and many other missions. He contributed to the design of Beam Wave Guide of 32m diameter antenna of the Indian deep space network station. Dr. Kumar is a Fellow of the Indian National Academy of Engineering (INAE) and a recipient of the "Young Scientist Merit Award" and the "Team Excellence Award" from ISRO; and the "Hari Ramji Toshniwal Award" and the "Prof. S. N. Mitra Memorial Award" from IETE, India. Low cross-pol microstrip antennas and array, spherical phased-array antennas, lightweight spacecraft antennas, microwave photonics, are some of his areas of special interest.

Sujoy Biswas is the Head of Electronics and Communication Engineering at the Department of Neotia Institute of Technology, Management and Science, India. He also served the Microwave Industry in India (2004–2007) as a Senior RF Design Engineer where he was engaged in design and development of various RF components for defense and space applications. He has made pioneering contributions in the field of DGS-integrated antennas and also co-authored a book chapter in this area titled "Defected Ground Structure for Microstrip Antennas" in *Microstrip and Printed Antennas: New Trends, Techniques and Applications*, Wiley, 2010. He has published numerous technical papers in international journals and conferences and also served different international symposia and conferences as a technical committee member. He is an active volunteer of IEEE and has served the local IEEE AP/MTT-S Chapter as a Chair (2020–2021).

Preface

This is an original and coveted document on DGS-based antennas presented for the first time encompassing its fundamentals to the state-of-the-art developments. The technique indeed is relatively new since "DGS," the abbreviated form of Defected Ground Structure, was primarily conceived in 2000 (Kim, Park, Ahn, Lim, *IEEE MWCL*, 10(4), 131–133, 2000) and that was quite focused on microwave printed circuit applications. The concept of applying DGS to moderate radiation and other properties of microstrip antennas dates back to 2005 (Guha, Biswas, Antar, *IEEE AWPL*, 4, 455-458, 2005), and we have been actively involved in initiating the concept as well as nurturing the subsequent developments over the decades.

DGS-based engineering immediately attracted various groups of antenna researchers including practicing engineers in the world's leading R&D laboratories. Their serious application-oriented approaches were quickly revealed through personal communications with us and through numerous articles widely published in journals and conference proceedings. A brief account of those was provided in the form of a book chapter in 2011 (Guha, Biswas, Antar, Ch. 12 in *Microstrip and Printed Antennas*, Ed. Guha & Antar, Wiley, 2011) as the first attempt to facilitate the engineers with more insights into theory and design.

It would be relevant to note that the first design of a DGS-integrated microstrip antenna, which we conceived in 2003 (reported in 2005 in IEEE AWPL), was completely based on a conjecture and approximate calculations without using any commercial simulator. Elementary versions of some simulation tools might have been available then in some limited advanced laboratories globally, but with no such traces in Indian academia. Some analytical models were known, but they were mostly geometry dependent.

Those limitations and challenges could not stop the antenna researchers; rather their enthusiasm added multiple colors to the study and useful applications to meet the need of the day. We have been contributing more toward the fundamentals in terms of analysis, design physics, and addressing the real challenges in application, and at the same time have been carefully watching the flare of the subject in various dimensions. We have been facing passionate technical queries starting from young researchers to veteran engineers, and parallelly receiving requests to write a complete book on this topic. This was the source of our motivation to start the ball rolling. Gradually we realized that it is difficult to write a complete book on a technology which is ever growing with new directions and innovations. That realization indeed forced us to wrap up the manuscript concentrating on the major needs and key requirements for both beginners like students and researchers, and R&D people working in industries.

The whole content in the book has been organized through nine chapters, and it is primarily based on our first-hand experience in dealing with the problems since the inception. We have tried our best to explain the physics and fundamentals of several intricate design aspects which are mostly untold in the open literature. At the same time, several published works have been addressed with relatively lesser in-depth discussions but with appropriate references so that the interested readers can quickly follow them up for detailed information.

The first two chapters are important to beginners. Chapter 1 gives a comprehensive introduction to the subject and to some extent a flavor of historical development of DGS. Chapter 2 is purely technical bearing all sorts of standard methods of theoretical analysis, modeling, and synthesis of DGS. Chapter 3 bridges the application of DGS from circuit to antennas. It discusses the concept of DGS underneath the printed feed lines to microstrip antennas to protect them from the higher harmonics commonly produced by the integrated amplifier or oscillator circuits. Chapters 4–6 address in-depth discussions and the most useful aspects of DGS-based techniques to control the cross-polarized (XP) radiations. Of these, Chapter 4 deals with the known physics behind XP generation in a standalone microstrip element and how DGS could be able to fight against the same. In contrast, Chapter 5 embodies more recent and advanced investigations enlightening the possible XP sources in microstrip radiators and also a few DGS-based solutions. Chapter 6 specifically addresses microstrip arrays and the challenges of DGS integration for improved performance. Apart from XP handling, another major potential of DGS is to mitigate mutual coupling commonly occurring between the microstrip array elements causing major shortcomings like radar scan blindness. Chapter 7 deals with DGS-based approaches to minimize the mutual coupling effect and hence improve the radiation issues. It covers the growing application of DGS in compact 5G MIMO and Millimeterwave antenna systems. Chapter 8 is a bit different from the rest of the chapters since it deals with circularly polarized (CP) patch design. This topic is relatively less investigated using DGS, although this chapter has unambiguously demonstrated the potential of DGS in addressing several challenges in achieving advanced CP performance. Chapter 9 embodies the application of DGS in UWB printed monopoles where the ground plane takes a major role in radiation. This chapter has shown how a

DGS helps in controlling the overall radiation of the antenna and also providing some essential narrow band notches within the ultrawide impedance bandwidth.

A beginner may consider this book as a guide to understand the subject and the research potential in this field. To a practicing engineer and an educator, we believe that this book would be a comprehensive source of up-to-date information and knowledge. Our efforts will be successful if our readers appreciate and find the book useful for them.

Debatosh Guha
Chandrakanta Kumar
Sujoy Biswas

Acknowledgments

Writing a book on a new and continuously developing technology is a rare experience. It offers enormous liberty and, at the same time, a profound responsibility. That is the reason why we started the process long ago and changed the format from time to time over the period of putting the materials together. The final organization of the manuscript has been distinctly appreciated with minor suggestions by some anonymous reviewers that indeed strengthened our efforts leading toward the final shape. We sincerely acknowledge them for providing quality time to review the manuscript and useful comments.

We express our thanks and indebtedness to our colleagues, co-researchers, coauthors, and students who have been associated with us and helped throughout the process. We would especially mention the names of our co-researchers Sk. Rafidul and Ms. Debi Dutta of the University of Calcutta, India, Dr. Chandreyee Sarkar of Birla Institute of Technology, Mesra, India, and Dr. Mohammad Intiyas Pasha of U R Rao Satellite Center, India, who helped tremendously in preparing the manuscript. Dr. Xiaoming Chen of Xi'an Jiaotong University, China, and Dr. Yan-Wen Zhao of University of Electronic Science and Technology of China, Chengdu, also helped by providing us with high-quality images from their published works. The publication process has been thoroughly guided and assisted by Ms. Aileen Storry, Wiley, Oxford, UK, and Ms. Kimberly Monroe-Hill, Wiley, New Jersey, USA. We are extremely thankful to both of them for their continuous help that has made our job easy.

We cannot refrain from acknowledging the ungrudging support and cooperation received from the members our family and also from our parent Institutions such as the Institute of Radio Physics and Electronics, University of Calcutta; the U R Rao Satellite Center, Bangalore; and the Neotia Institute of Technology Management and Science, Kolkata, India.

1

Introduction to DGS: The Concept and Evolution

1.1 Introduction

The last decade has witnessed significant advancements in the area of wireless communication with the advent of new technologies like 4th (4G) and 5th (5G) generation mobile communication, massive Multiple-Input Multiple-Output (MIMO) systems, and many others. The possibility of Terahertz radio communication is also being explored for the future. The primary goal of such continuous innovation is to develop fast and reliable communication systems capable of handling the exponentially growing data traffic. Realization of such complex and advanced technologies require energy efficient compact devices. The primary challenge is to develop the compatible hardware. The radio frequency (RF) front ends are the most important components of that. Various novel techniques have been explored to meet the hardware requirements as well as in improving the performance of printed circuits and antennas.

Defected Ground Structure (DGS) has been one of them since its inception in 2000. Within a short period of time, it has been successfully applied to a wide variety of miniaturized RF components and antennas. Especially, the application of DGS in designing planar antennas witnessed immense growth in the last decade. A sprawling number of technical papers, patents, and book chapters have been produced in a span of the last 15 years. A few of these techniques have already been adopted in commercial products. Such popular applications and success indicate inherent simplicity of DGS-based design along with its commercial viability.

This book exclusively deals with DGS integration techniques for antenna applications and is the first of its kind. Therefore, the introductory chapter has been designed to address some insightful fundamentals which cover the evolution starting from the concept of photonic crystals, the working principles, known

Defected Ground Structure (DGS) Based Antennas: Design Physics, Engineering, and Applications,
First Edition. Debatosh Guha, Chandrakanta Kumar, and Sujoy Biswas.
© 2023 The Institute of Electrical and Electronics Engineers, Inc. Published 2023 by John Wiley & Sons, Inc.

geometries along with an account of their chronological developments, and different areas of applications.

1.2 Evolution of DGS

A reader must be curious to know, what is DGS? Before addressing this question, one needs to understand the fundamental concept initially developed in the domain of optics. Conventional materials that are abundantly available in nature have been the primary resources to the engineers. But the technologists explore new engineered materials to solve several problems. The pioneering study dates back to 1987 [1] when Yablonovitch created a three-dimensional periodic crystal by drilling holes through a dielectric block in the form of a diamond lattice and observed the restricted propagation of light, i.e. electromagnetic (EM) waves in a three-dimensional space compared to that occurring in all directions through a traditional dielectric block. The situation is analogous to the propagation of electrons as waves in crystalline semiconductor materials [2]. Such artificially engineered crystals are also able to restrict the wave propagation over a certain band of wavelengths or polarizations. They were termed as Photonic Band-Gap (PBG) materials or photonic crystals [3, 4].

The term "photonic band-gap" and its concept were adopted from optics and solid state physics. The microwave engineers explored analogous periodic structures in various forms to control the transmission of EM waves by producing stopband characteristics. Initially, these geometries were commonly termed as PBG structures, but soon a debate arose [5]. The microwave groups preferred to call them as "Electromagnetic Band-gap (EBG)" materials [6, 7]. A representative diagram indicating a stopband in transmission characteristics caused by an EBG structure is shown in Figure 1.1. Such structures at microwave frequencies would be of periodic nature and they can be realized by an arrangement of simple dielectric units [8, 9], or composite metallo-dielectric units [10], or metallic conductors [11] with a lattice period $p = n\lambda_g/2$, λ_g being the guide wavelength. According to the periodicity, they are usually classified into three categories: (i) three-dimensional periodic structure [8, 9], (ii) two-dimensional periodic structure [12–15], and (iii) one-dimensional structure [16, 17]. An example is depicted through Figure 1.2. Some realistic EBG structures of various geometries and periodicity are shown in Figure 1.3.

The EBG structures, therefore, were widely considered for microwave filtering applications [12, 18, 19]. A few more circuit applications also became popular among the microwave engineers. The antenna engineers also started exploring EBG toward some antenna related problems which included reduction of surface waves and mutual coupling issues in phased arrays [20, 21], and

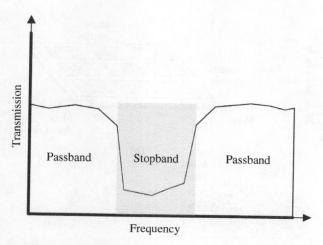

Figure 1.1 Transmission characteristics of a typical EBG structure indicating stop and passbands.

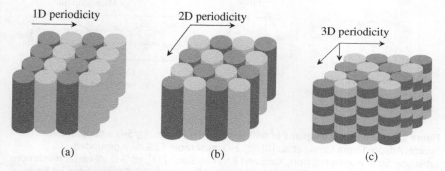

Figure 1.2 Representative diagram indicating the possible periodicity in EBG structures: (a) 1D periodicity, (b) 2D periodicity, (c) 3D periodicity.

increasing antenna gain [22, 23]. One such example is shown in Figure 1.4 where a microstrip antenna integrated with a 3D EBG structure results in enhanced gain [22]. The rapid advancement in EBG research also made some interesting applications to Global Positioning System (GPS), bluetooth, mobile, and wearable antenna designs [6, 24, 25]. In principle, an EBG structure could be easily three dimensional but for printed microwave circuits or transmission lines, 2D planar geometries are more logical and user friendly. The ground planes appear to be the best possible areas to implement them as printed patterns. They appear much easier and simpler compared to perturbing the substrate or drilling holes through it. One such 2D planar EBG is shown in Figure 1.5 [13]. This actually bears

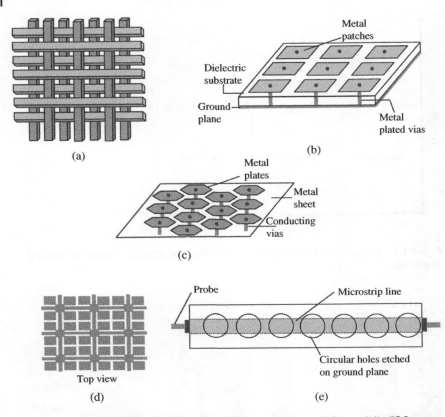

Figure 1.3 Schematic diagrams of different EBG structures: (a) 3-D woodpile EBG. Source: Adapted from Ozbay et al. [8]; (b) 2-D mushroom EBG on a grounded substrate. Source: Adapted from Yang and Rahmat-Samii [22]; (c) 2-D all metal mushroom EBG. Source: Adapted from Sievenpiper [15]; (d) 2D printed EBG. Source: Adapted from Yang et al. [14]; (e) 1-D printed EBG on the ground plane. Source: Refs. [16, 17].

Figure 1.4 A circularly polarized curl antenna placed on an EBG structure for gain enhancement. Source: Yang et al. [22] John Wiley & Sons, Inc.

multiple unit cells etched out on the ground plane and arranged in a periodic fashion beneath a microstrip line.

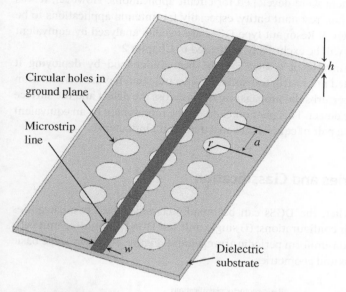

Figure 1.5 Periodic two-dimensional (2D) circular defects pattern etched on the ground plane beneath a microstrip transmission line. Source: Adapted from Radisic et al. [13].

In 1999, a group of researchers [26] further simplified the geometry and discarded the periodic nature of the pattern. They simply used a unit cell to achieve considerable stopbands in C and X-band microstrip line and called it "PBG unit structure" in their introductory article [26]. The name is aptly justified by the geometrical shape of the unit cell and will be discussed in details in the following section. In a subsequent article [27], the same structure was termed as *Defected Ground Structure* or its abbreviation "DGS."

Therefore, a DGS may be regarded as a simplified variant of printed EBG on the ground plane.

1.3 Definition and Basic Concept

DGS means a single or a limited number of slots strategically etched on the ground plane of a microwave printed circuit board (M-PCB) to attribute a feature of stopping wave propagation over a band of frequencies. The slot or "defect" is actually a compact geometry, commonly known as a "unit cell." A DGS thus can be described as a planar EBG with unit cell or a periodic arrangement of a limited number of cells. The shape and size of the unit cells determine their resonant characteristics and most of them were developed for circuit applications. However, it was also explored as a non-resonant entity, especially for antenna applications to be discussed in Chapter 4. Resonant type DGSs are usually analyzed by equivalent circuit models and will be exclusively addressed in Chapter 2.

The working principle of a DGS can be best understood by deploying it underneath a printed microstrip transmission line and studying the transmission characteristics. It perturbs the propagating fields across the defect and the surface current around the defect. This phenomenon can be represented by an equivalent circuit comprising a pair of equivalent capacitance and inductance.

1.4 Geometries and Classification

As mentioned earlier, the DGSs can be broadly classified into two categories depending on their configurations: (i) single unit cell DGS (ii) multiple unit cells with uniform or non-uniform periodic arrangement. Figure 1.6 shows some basic DGS classifications and geometries.

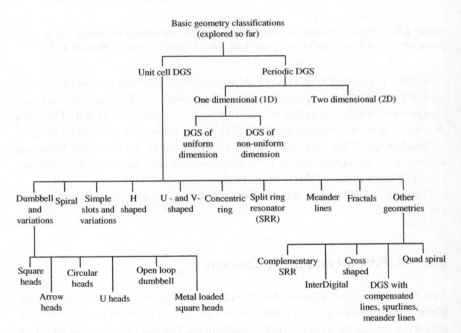

Figure 1.6 DGS: basic classifications and geometries.

1.4.1 Unit Cell DGS

In the last two decades, numerous DGS geometries have been investigated and reported in open literature for different applications. Some of them are inspired from natural geometrical shapes that we experience in our daily life, while others are more complex in nature and as such cannot be specifically named. These shapes have been explored with an aim to achieve improved performance in terms of stop or passband characteristics, compactness, and ease of design based on varied requirements. The known geometries, reported so far, include simple shapes such as rectangular dumbbell [27], circular dumbbell [28], spiral [29], "U" [30], "V" [30], "H" [31], "T" [32], "E" [33], cross [34], concentric rings [35], skewed "F" [36]. Complex DGS shapes follow the geometries of split ring resonators (SRRs) [37, 38], fractals [39], Minkowski couple [40], Hilbert-shaped complimentary SRR (CSRR) [41], etc. Some of them are depicted through Figure 1.7. Most of the said DGS shapes were explored to implementing band stop filters, suppressing the surface waves, rejecting harmonics, compacting microwave circuits, impedance matching, etc.

1.4.1.1 Dumbbell-Shaped DGS

It is the first unit cell DGS which was reported by Park et al. [26] and subsequently by Kim et al. [27]. The DGS looks like a "dumbbell" and as such it is popularly called "dumbbell shaped DGS." It consists of two rectangular slots connected by a narrow slot etched on the ground plane as shown in Figure 1.8a. The S-parameters of a 50 Ω microstrip line passing over that DGS are shown in Figure 1.8b. As discussed above, a distinct stopband property is evident in the transmission characteristics as a function of DGS dimensions. For a better visualization, the same structure [27] has been simulated using [42] and the visual portrays are depicted through Figure 1.9. The ground plane current (Figure 1.9a,b) is found to go around the DGS square heads to form a current loop and hence an equivalent inductance "L" may be used to model the effect. The trapped E-fields across the gap (Figure 1.9c) produce an equivalent capacitance "C." Thus, a resonant trap is created across the DGS which works over a band of frequencies around the resonance. This is revealed as a stopband in the transmission characteristics of the line. Apart from this, the modified line parameters also alter the line impedance. This feature is conveniently employed in many applications is to realize higher impedances. Therefore, a DGS beneath a transmission line can reveal the following features: (i) stopband properties like EBG structures, (ii) realization of high impedance line, and (iii) slow wave effect due to long current path around the DGS. The quantitative analysis will be introduced in Chapter 2.

Figure 1.9e,f portrays the simulated E-fields captured over the pass and stopband frequencies respectively. It was observed earlier that the stopband is a strong function of DGS dimensions (Figure 1.8b). This in turn indicates a change in its equivalent inductance and capacitance with the defect dimensions. The quantitative analysis with more insight into this phenomenon will be discussed in Chapter 2.

1.4.1.2 Variations of Dumbbell-Shaped DGS

The dumbbell-shaped DGS [27] and its subsequent applications to filter design [43] had a huge impact on microwave circuit design with some special foci on implementation area, stopband property, insertion loss, etc. [28, 44–46]. A comprehensive overview is presented through Figure 1.7. These different shapes of the slots indeed control the resonant frequencies of the DGS and thus the stopbands. Circular head dumbbell DGS is more compact relative to rectangular head. The arrowhead DGS (e.g. Figure 1.7e) produces sharper stopband transition [28] compared to the circular head version. A detailed discussion is available in Chapter 2 which deals with the DGS characteristics in terms of equivalent circuit model.

Compact DGS has been an all-time priority for any application. But at the lower frequencies, the defect dimension gets larger. Such an issue has been intelligently handled in [47] by reshaping the defect. This is shown in Figure 1.10a which indeed enhances the resultant capacitance value without altering the inductance. That helps in pushing the resonance frequency toward the lower side of spectrum. The geometry of Figure 1.10a may also be thought of as metal loaded dumbbell and a similar application was tried for coplanar waveguide (CPW) [48]. Further alterations were made like double loading as shown in Figure 1.10b [48] to achieve even lower stopband frequencies.

Figure 1.7 Different DGS geometries: (a) Dumbbell-shaped. Source: Adapted from Ref. [27] (b) Spiral-shaped. Source: Adapted from Ref. [29] (c) H-shaped. Source: Adapted from Ref. [31] (d) U-shaped. Source: Adapted from Ref. [30] (e) Arrow head dumbbell. Source: Adapted from Ref. [28] (f) Concentric ring shaped. Source: Adapted from Ref. [35] (g) Complimentary Split-ring resonators. Source: Adapted from Ref. [37] (h) Interdigital. Source: Adapted from Ref. [69] (i) Cross-shaped. Source: Adapted from Ref. [34] (j) Circular head dumbbell. Source: Adapted from Ref. [28] (k) Square heads connected with U slots. Source: Adapted from Ref. [46] (l) Open loop Dumbbell. Source: Adapted from Ref. [50] (m) Fractal. Source: Adapted from Ref. [39] (n) coupled Half-Circles. Source: Adapted from Ref. [49] (o) V-shaped. Source: Adapted from Ref. [30] (p) L-shaped. Source: Adapted from Ref. [71] (q) Meander Lines. Source: Adapted from Ref. [71] (r) U-head dumbbell. Source: Adapted from Ref. [51] (s) double equilateral U. Source: Adapted from Ref. [62] (t) Square slots connected with narrow slot at edge. Source: Adapted from Ref. [46] (u) quad-spiral DGS. Source: Adapted from Ref. [59] (v) open square. Source: Adapted from Ref. [70] (w) T-shaped. Source: Adapted from Ref. [32] (x) E-shaped. Source: Adapted from Ref. [33] (y) CSRR-head dumbbell. Source: Adapted from Ref. [53]. Solid color means "defect" on the ground plane.

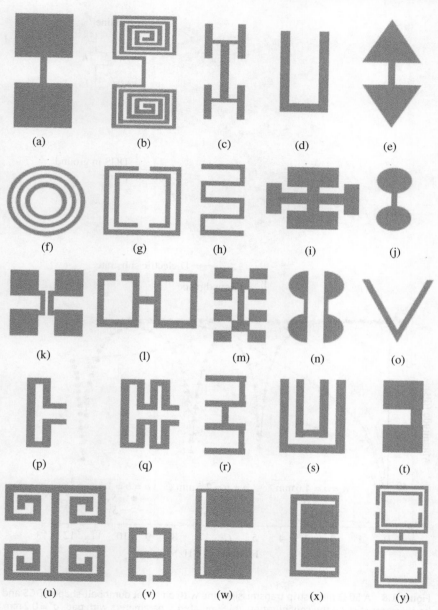

(a) (b) (c) (d) (e)

(f) (g) (h) (i) (j)

(k) (l) (m) (n) (o)

(p) (q) (r) (s) (t)

(u) (v) (w) (x) (y)

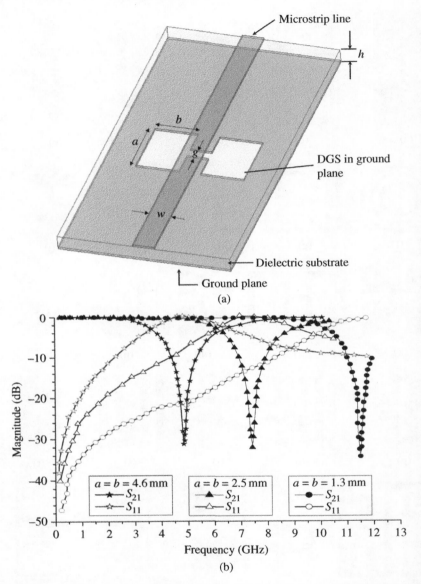

(a)

(b)

Figure 1.8 A 50 Ω microstrip transmission line with a typical dumbbell-shaped DGS and its characteristics (a) the configuration; (b) Simulated S-parameters with gap "g" = 0.2 mm. TACONIC substrate with $\varepsilon_r = 10$ and thickness 62 mil. Source: Kim et al. [27] © [2000] IEEE.

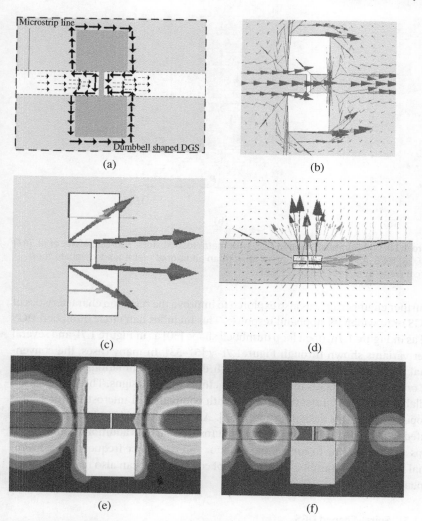

Figure 1.9 (a) Schematic diagram of ground plane current around the defect; (b) simulated portray of the same ground plane current around the defect; (c) Trapped E-field across the defect; (d) H-fields due to the current loop round the square dumbbell heads; (e) Simulated E-field distribution at the passband frequency (2 GHz); (f) Simulated E-field distribution at stopband frequency (8 GHz).

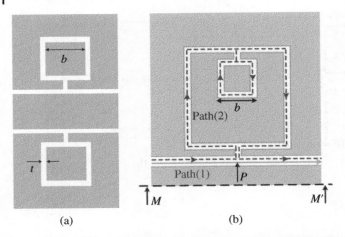

(a)   (b)

Figure 1.10 (a) Dumbbell shaped DGS with metal loaded slots. Source: Safwat et al. [47] © [2006] IEEE; (b) Modified version of (a) with an additional metal loaded square loop. Source: Wang et al. [48] © [2012] IEEE.

Further engineering was also explored to improve the rejection characteristics of a DGS or to widen its stopband response. This includes half circle dumbbell DGS [49] as in Figure 1.7n, open loop dumbbell shape [50] as in Figure 1.7l, and several other variants shown through Figure1.7r–y [51–53]. In many cases, the conventional dumbbell DGS has been used with different combinations of microstrip line or parasitic elements to obtain compactness in the designs. The examples are available in [54, 55]. Integrating DGSs with compensated microstrip line is also a popular choice for designing LPFs [56]. An interesting idea has been demonstrated in [57] where a DGS integrated microstrip line uses additional small metal strips around the line to control the Q-factor and resonance frequency. A conventional square head dumbbell with a pair of coupled slots can also result in a wide stopband LPF [58].

1.4.1.3 Spiral-Shaped DGS

Dumbbell-shaped DGSs were successfully implemented in low-pass filters (LPFs) design. Further improvements were also in search with the help of newer geometries. Since the defect dimensions are directly related to the operating wavelength, their focus was on achieving compact DGS for lower frequency designs. Thus, a spiral-shaped DGS, as shown in Figure 1.11, was conceived and reported in [29]. Comparing a spiral with a dumbbell shape occupying identical surface area on an identical substrate, it is apparent that the attenuation pole for a spiral DGS occurs at much lower frequency relative to that due to the dumbbell DGS. Increased equivalent inductance L and capacitance C in a spiral structure are the primary reason behind this. This indeed implies relative compactness of a spiral head DGS which also produces higher Q factor resulting in relatively steeper rejection characteristics.

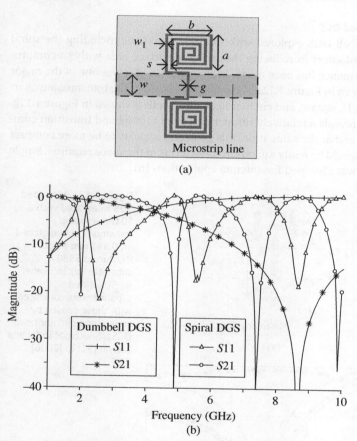

(a)

(b)

Figure 1.11 Top view of spiral shaped DGS integrated with 50 Ω microstrip transmission line fabricated on a 31 mil thick RT/Duroid 5880 substrate: (a) schematic diagram; (b) Comparison of simulated *S*-parameters with an identical line on dumbbell-shaped DGS. Source: Kim et al. [29] with permission from The Institution of Engineering and Technology.

1.4.1.4 Variations of Spiral-Shaped DGS

An interesting variation of spiral DGS was explored to design an attenuator [59]. This uses a pair of spiral heads on each side and therefore a total of four heads in the structure as shown in Figure 1.7u. This is called quad spiral DGS which provides further lowering in operating frequencies due to the increased equivalent inductance. Like dumbbell DGS, spiral DGSs have also been designed for CPWs [60]. The other variations of spiral geometry will be addressed in a following section entitled "asymmetric DGS."

1.4.1.5 H-Shaped DGS

After the dumbbell-DGS explored with multiple variations including the spiral heads, a constant effort in reducing the required surface area without compromising in performance has been made. H-shaped DGS [31] was one of the major outcomes as shown in Figure 1.12a. A comparative study of the transmission characteristics using H, square, and circular head dumbbells is shown in Figure 1.12b. H-shaped DGS reveals a relatively sharper passband to stopband transition compared to others and at the same time, this geometry appears to be more compact in nature. This would be really a good choice in terms of the space required. Single H-shaped DGS was also used for antenna applications [61].

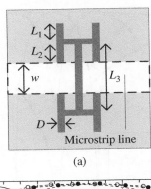

(a)

Figure 1.12 H-shaped DGS integrated with a 50 Ω microstrip transmission line etched on 0.381 mm thick RT/Duroid 5880 substrate: (a) Top view; (b) Simulated S-parameters compared with those caused by square and circular head DGSs. Source: Mandal and Sanyal [31] © [2006] IEEE.

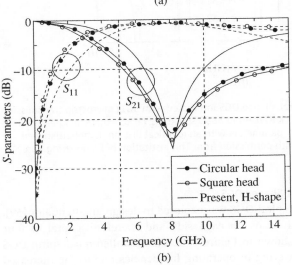

(b)

1.4.1.6 U- and V-Shaped DGSs

The DGS shapes studied in the early 2000s were applied mostly to filter designs. However, the requirement of filters is different for varied applications. Sometimes a filter with a very sharp and deep band rejection is required for suppressing spurious signals. Apart from H-shape, U- and V-shaped DGSs were successfully examined for that purpose [30]. These geometries are shown in Figure 1.7d and o. A U-shaped DGS and a spiral DGSs with identical resonance frequency exhibit a noticeable difference in their Q values. The separation between the arms of a U-DGS or the angle subtended by the arms of a V-DGS also controls their Q. If a typical U-DGS shows $Q \approx 36$, its corresponding version of spiral DGS gives only 7.5 [30] and thus relatively sharp rejection characteristics are revealed by a U-DGS. Different variations of U shapes have been explored and one such example is shown in Figure 1.7s [62]. This is capable of producing dual stopbands independently as a function of the slot parameters. Another variant [32] is depicted in Figure 1.13 which claims higher Q with compact geometry.

Figure 1.13 U-shaped DGS: (a) Conventional type [30]; (b) Modified geometry with extra slots. Source: Wang et al. [32] © [2007] IEEE.

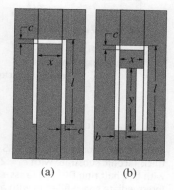

(a)   (b)

1.4.1.7 Ring-Shaped DGS

A completely different kind of geometrical shape was proposed in [35]. The geometry looks like Figure 1.7f and as such was termed as "concentric ring DGS." This comprises concentric circular rings etched out on the ground plane and was found to reveal a considerably wide stopband. The configurations in Figure 1.14a,b were examined in [35] revealing their bandstop properties as a function of the ring radii cum their width. That study [35] also demonstrated a metal backing beneath the DGS as a remedial measure to any leakage of RF power through the defects as back radiations. The scheme and the outcome are shown in Figure 1.14c,d, respectively. The concept of ring DGS [35] was customized to obtain different variations for applications to the antenna engineering. They include single ring DGS [63, 64], arc-shaped DGS [64, 65], open-ring DGS [66, 67], and truncated ring DGS [68]. Detailed discussions will appear in Chapters 3, 4 and 7.

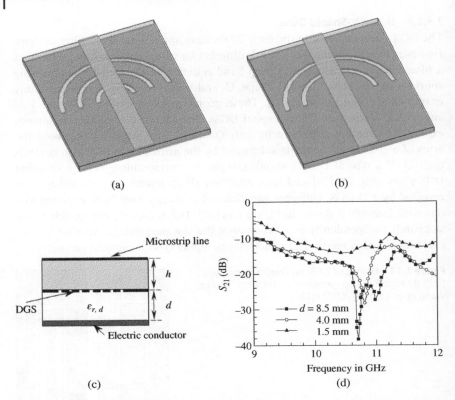

Figure 1.14 Concentric ring DGS beneath a 50 Ω microstrip line etched on a 1.575 mm thick Taconic substrate with $\varepsilon_r = 2.2$: (a) Top view with three half ring DGS; (b) Top view with two half ring DGS; (c) Cross-sectional view when backed by a metallic sheet and the intermediate space filled in with a dielectric medium; (d) Measured transmission characteristics as a function of variable spacing d. Source: Guha et al. [35] © [2006] IEEE.

1.4.1.8 Other DGS Geometries

A few important DGS geometries are discussed above and indeed it has no limits. Numerous shapes have been explored so far demonstrating attractive features and applications. The relevant discussions are provided in Chapters 4 and 6. The geometry shown in Figure 1.7i looks like a cross which is actually a superposition of H- and I-shaped DGSs. The interdigital slots in Figure 1.7h [69] enables one to vary the resonance frequency by adjusting the finger lengths without changing the occupied area. Steeper passband to stopband transition is possible by SRR shaped

DGSs [37, 38]. This is a good candidate for filter design with elliptic response. Quasi elliptic filter function can also be achieved using open square DGS shown in Figure 1.7v [70]. Multi stopband feature may be of interest in many circuit applications for which L-shaped defect, shown in Figure 1.7p [71], would work well.

Another interesting high Q geometry is shown in Figure 1.15a which is called hairpin DGS [72]. The geometry is a bit complex although more complex shapes have also been reported [39] and listed in Figure 1.7. A configuration like a Hilbert curve ring (HCR) [41] is shown in Figure 1.15b. This HCR-DGS is useful for designing LPFs with ultra-wide stopbands and low insertion loss.

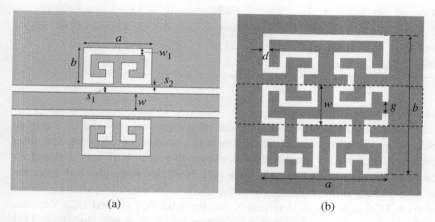

(a)   (b)

Figure 1.15 DGS with special shapes: (a) Hairpin shaped DGS. Source: Lee et al. [72] John Wiley & Sons; (b) Hilbert curve ring SRR-shaped DGS. Source: Xu et al. [41] with permission from Elsevier.

A gross comparative study is presented in Table 1.1 for a set of popular DGS geometries. This would give an idea about the relative dimensions along with the required deployment area, Q values, and percent bandwidth. The spiral geometry appears to be the best one in terms of compactness and reasonably high Q. The E-shaped DGS could be its close competitor in terms of the deployment area. But V- and U- DGSs are far above the others in terms of the Q values. They are ideal for designing the notch filters. The hairpin DGS is with moderately high Q and compactness. The ultimate choice depends on the technical requirements of a specific application since all DGS geometries have evolved on that basis. A few examples are circular dots [73], linear slot [74], arc or bracket like slots [75, 76], hexagonal aperture [77], etc. These are discussed in Chapters 4 and 6.

Table 1.1 Performance of some selected popular DGS shapes.

Geometry	Center frequency f_0 (GHz)	% Bandwidth	Q factor	Implementation area/λ_g^2
Dumbbell [27]	4.8	62.5	1.6	0.125
Spiral [29]	2	14	7.14	0.0062
U shaped [30]	3.36	6.5	15.27	0.011
V shaped [30]	4.96	3.6	27.55	0.06
H shaped [31]	5	60	1.66	0.025
T shaped [32]	3.91	32.22	3.1	0.029
E shaped [33]	5	18.4	5.43	0.0067
SRR [38]	3.8	20.26	4.93	0.01
Hairpin [72]	4.68	8.33	12	0.08

1.4.1.9 Tunable DGS Geometries

The feature of tunability of resonance has been examined with the DGS geometries. Such flexibility should be of much importance in adjusting the filter response. Patch loaded dumbbell DGS, shown in Figure 1.10, was explored for the first time as a reconfigurable filter [47]. It allows some physical space for accommodating varactor diode where the diode capacitance in parallel with the inherent DGS capacitance determines the stopband. A T-shaped DGS, depicted in Figure 1.7w, was made tunable by integrating with a varactor diode [32]. A scheme is shown in Figure 1.16 [32] which features a tuning range of 13.2% around 2.35 GHz.

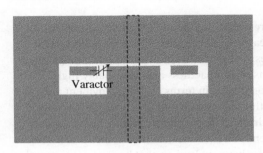

Figure 1.16 Modified T-shaped defect bearing a varactor diode. Source: Wang et al. [32] © [2007] IEEE.

1.4.2 Periodic Uniform DGS

Repetition of *unit cells* in a periodic fashion sometimes helps in achieving deeper, steeper and wider stopband characteristics. A DGS thus formed by repetition of unit cells is referred to as *Periodic DGS*. Figure 1.17a [78] shows a five-cell periodic DGS fabricated on Taconic substrate with a 50 Ω microstrip line printed on its other side. Its transmission characteristics, presented in Figure 1.17b, show much wider stopband compared to that produced by a single identical cell. This also indicates that the stopband cutoff depends on the cell dimensions. The period marginally affects the center of the band. Rather, the number of cells determines the width and rejection depth of S_{21} versus frequency plot.

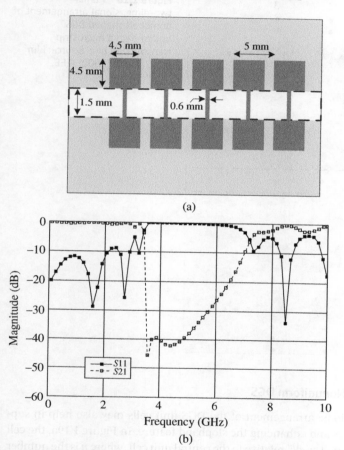

Figure 1.17 One-dimensional 5-cell uniform dumbbell-shaped DGS beneath a 50 Ω microstrip line fabricated on a 1.575 mm thick TACONIC-CER 10 substrate with $\varepsilon_r = 10$: (a) schematic view; (b) Measured S-parameters. Source: Liu et al. [78] © [2004] IEEE.

Most of the periodic DGS uses 1D arrangement of unit cells. Some applications may need 2D periods. One interesting example is shown in Figure 1.18 [79], where a number of cells are periodically arranged in a vertical column and then each column is periodically arranged along the horizontal direction. This study was mainly focused to achieve increased slow-wave factor for microstrip and CPW lines compared to that achievable using EBG structures or 1D periodic DGS. The increased slow wave factor is effectively used to reduce the size and compacting microwave circuits [79].

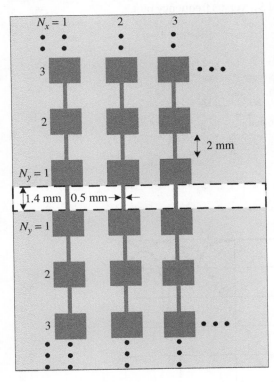

Figure 1.18 Planar two-dimensional arrangement of dumbbell shaped DGS underneath a microstrip transmission line. Source: Lim et al. [79] © [2002] IEEE.

1.4.3 Periodic Nonuniform DGS

Non uniform periodic arrangement of the DGS unit cells may also help in suppressing the ripples and enhancing the stopband feature. In Figure 1.19a, the cell dimensions are varied as $e^{1/n}$ relative to the central unit cell, where n is the number of the elements on either side including the central one. Figure 1.19a works with $n = 3$ and the amplitude distribution of the units are [78]:

$$e^{1\backslash 3}\, e^{1\backslash 2}\, e^1\, e^{1\backslash 2}\, e^{1\backslash 3} \approx 1.396, 1.649, 2.718, 1.649, 1.396.$$

Here, the central dumbbell, chosen to have its amplitude "e^1" has 4.5 mm side length. Then the other dumbbell cells have side lengths 2.7 and 2.5 mm following

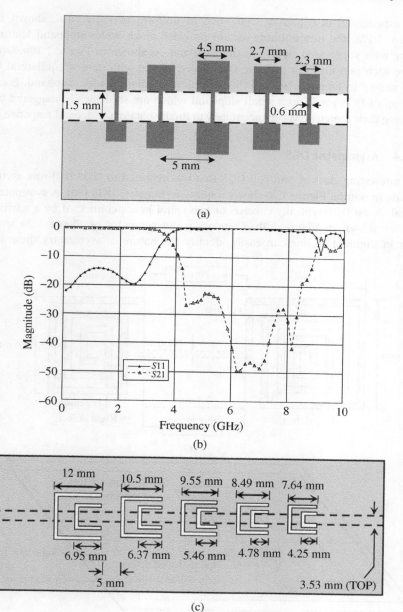

(a)

(b)

(c)

Figure 1.19 One-dimensional 5-cell non-uniform periodic DGS: (a) Top view for non-uniform dumbbell-shaped DGS. Source: Adapted from Liu et al. [78]; (b) Measured S-parameters for the structure shown in (a); (c) Top view for non-uniform double equilateral U shaped DGS. Source: Adapted from Ting et al. [62].

the exponential distribution. Compared to uniform periodic DGS, shown in Figure 1.17a, the non-uniform variant provides much wider stopband feature along with suppressed ripples in the passband, as shown in Figure 1.19b. One more such non uniform periodic DGS was explored with double equilateral U [62] shown in Figure 1.19c and this was used to obtain a wide stopband. Each section of DGS produces a small stopband which are strategically staggered by varying their dimensions very accurately to finally obtain a wideband response.

1.4.4 Asymmetric DGS

An interesting class of unit cell DGS that was reported in 2005 [80] was asymmetric in nature. Figure 1.20 shows a spiral asymmetric DGS [80]. A symmetric spiral shape DGS typically consists of two spiral heads connected by a narrow slot as shown in Figure 1.7b. Comparing the geometry of Figure 1.20 with that in Figure 1.7b one can easily identify the nature of asymmetry there in.

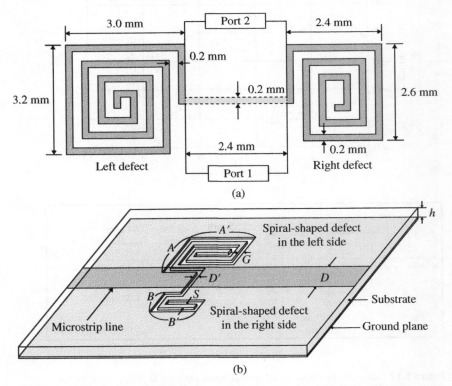

Figure 1.20 Asymmetric spiral DGS : (a) the topology; (b) 3D schematic view when integrated with a 50 Ω microstrip line. Source: Woo and Lee [80] © [2005] IEEE.

The dimension of one of the spiral heads is smaller than the other. This helps in achieving dual resonances due to the different size of the spiral heads. This asymmetric DGS is used for Wilkinson power divider to suppress the second and third harmonics with simultaneous reduction in size. Yet another asymmetric DGS employing conventional dumbbell-shaped configuration was studied in [81] with elliptic response shown in Figure 1.21a. LPF designed with such a DGS exhibited sharp passband to stopband transition along with low insertion loss and broader stopband compared to conventional dumbbell-shaped DGS. Similar structure was tried with metal loading for lowering the operating frequency [82] as shown in Figure 1.21b.

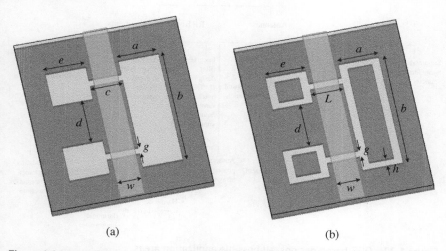

(a)   (b)

Figure 1.21 Asymmetric dumbbell-shaped DGS: (a) simple slotted geometry. Source: Adapted from Parui and Das [81]; metallic patch loaded configuration. Source: Adapted from Moyra et al. [82].

1.5 An Outline of Applications

Since its inception in 2000, the DGS has been applied to almost all major printed circuits and antennas to enhance their performance and mitigate some design issues. A chart shown in Figure 1.22 portrays a systematic overview of its application areas. They may be broadly classified into two categories, such as (i) RF front ends and circuits and (ii) Antennas. In its early stages of development, DGSs were applied to designing the LPFs [43]. The potential of this technique was soon understood and parallelly applied to a wide variety of high-performance bandpass, band stop, low-pass, and reconfigurable filters [83, 84]. The last decade has witnessed rapid developments and applications of DGS. Apart from filters, it found applications to power amplifiers [85–88], oscillators [89–91], mixers [92, 93], couplers [45, 94, 95], power dividers [96], frequency doublers [97], diplexers [98], phase shifters [99], etc. Very recently, DGS has been explored in wireless power transfer applications [100, 101]. DGS-based antenna was first conceived in 2005 [73] to improve the radiation characteristics of microstrip patch radiators. Before [73], DGS was explored for printed antennas, but as a filter beneath the feed lines to reject the higher harmonics of the input signal.

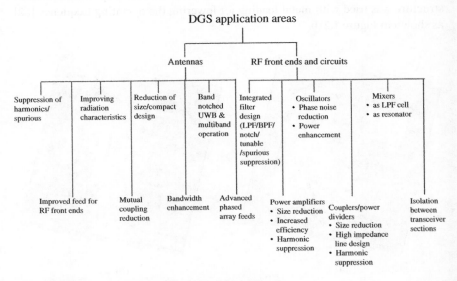

Figure 1.22 DGS-based designs: all possible application areas.

After [73], the DGS-based antenna design evolved as a new branch of research as well as applications [83, 84]. They are mostly oriented around printed antennas, arrays, and to limited cases of Dielectric Resonator Antennas. The rest of the chapters will primarily focus on the antenna aspects covering analysis, design physics and insight, along with the engineering challenges.

References

1 E. Yablonovitch, "Inhibited spontaneous emission in solid-state physics and electronics," *Physical Review Letters*, vol. 58, no. 20, pp. 2059–2062, 1987.

2 S. John, "Strong localization of photons in certain disordered dielectric superlattices," *Physical Review Letters*," vol. 58, no. 23, pp. 2486–2489, 1987.

3 E. Yablonovitch, "Photonic band-gap structures," *Journal of the Optical Society of America A,* vol. 10, no. 2, p. 283, 1993.

4 J. D. Joannopoulos, R. D. Meade, and J. N. Winn, *"Photonic Crystals," Princeton U. Press, New Jersey, U.S.A.* 1995.

5 A. A. Oliner, "Periodic structures and photonic-band-gap terminology: historical perspectives," *29th European Microwave Conference,* pp. 295–298, 1999, Munich, Germany.

6 P. de Maagt, R. Gonzalo, Y. C. Vardaxoglou, and J. M. Baracco, "Electromagnetic bandgap antennas and components for microwave and (sub)millimeter wave applications," *IEEE Transactions on Antennas and Propagation,* vol. 51, no. 10, pp. 2667–2677, 2003.

7 F. Yang and Y. R. Samii, "Electromagnetic Bandgap Structures in Antenna Engineering" *Cambridge University Press,* 2008.

8 E. Ozbay, A. Abeyta, G. Tuttle, M. Tringides, R. Biswas, T. Chan, C. M. Soukoulis, and K. M. Ho, "Measurement of a three-dimensional photonic band gap in a crystal structure made of dielectric rods," *Physical Review B: Condensed Matter,* vol. 50, no. 3, pp. 1945–1948, 1994.

9 K. M. Ho, C. T. Chan, C. Soukoulis, R. Biswas, and M. Sigalas, "Photonic band gaps in three dimensions: new layer-by-layer periodic structure," *Solid State Communications,* vol. 89, no. 5, pp. 413–416, 1994.

10 H. Contopanagos, L. Zhang, and N. G. Alexopoulos, "Thin frequency-selective lattices integrated in novel compact MIC, MMIC, and PCA architectures," *IEEE Transactions on Microwave Theory and Techniques,* vol. MTT-46, no. 11, pp. 1936–1948, 1998.

11 A. S. Barlevy and Y. Rahmat-Samii, "Characterization of electromagnetic band-gaps composed of multiple periodic tripods with interconnecting vias: concept analysis, and design," *IEEE Transactions on Antennas and Propagation,* vol. 49, no. 3, pp. 242–353, 2001.

12 I. Rumsey, M. Piket-May, and P. K. Kelly, "Photonic bandgap structure used as filters in microstrip circuits," *IEEE Microwave and Guided Wave Letters,* vol. 8, no. 10, pp. 336–339, 1998.

13 V. Radisic, Y. Qian, R. Coccioli, and T. Itoh, "Novel 2-D photonic bandgap structure for microstrip lines," *IEEE Microwave and Guided Wave Letters,* vol. 8, no. 2, pp. 69–71, 1998.

14 F.-R. Yang, K.-P. Ma, Y. Qian, and T. Itoh, "A uniplanar compact photonic-bandgap (UC-PBG) structure and its applications for microwave circuit," *IEEE Transactions on Microwave Theory and Techniques,* vol. 47, no. 8, pp. 1509–1514, 1999.

15 D. Sievenpiper, L. Zhang, R. F. J. Broas, N. G. Alexopolus, and E. Yablonovitch, "High-impedance electromagnetic surfaces with a forbidden frequency band," *IEEE Transactions on Microwave Theory and Techniques*, vol. 47, no. 11, pp. 2059–2074, 1999.

16 F. Falcone, T. Lopetegi, and M. Sorolla, "1-D and 2-D photonic bandgap microstrip structures," *Microwave and Optical Technology Letters*, vol. 22, no. 6, pp. 411–412, 1999.

17 C. C. Chiau, X. Chen and C. Parini, "Multiperiod EBG structure for wide stopband circuits," *IEE Proceedings – Microwave, Antennas and Propagation*, vol. 150, no. 6, pp. 489–492, 2003.

18 N. Yang, Z. N. Chen, Y. Y. Wang, and M. Y. W. Chia, "A two-layer compact electromagnetic bandgap (EBG) structure and its applications in microstrip filter design," *Microwave and Optical Technology Letters*, vol. 37, no. 1, pp. 62–64, 2003.

19 L. Yang, M. Fan, F. Chen, J. She, and Z. Feng, "A novel compact electromagnetic-bandgap (EBG) structure and its applications for microwave circuits," *IEEE Transactions on Microwave Theory and Techniques*, vol. 53, no. 1, pp. 183–190, 2005.

20 R. Gonzalo, P. de Maagt, and M. Sorolla, "Enhanced patch antenna performance by suppressing surface waves using photonic band-gap structures," *IEEE Transactions on Microwave Theory and Techniques*, vol. 47, no. 11, pp. 2131–2138, 1999.

21 F. Yang and Y. Rahmat-Samii, "Microstrip antennas integrated with electromagnetic band-gap (EBG) structures: a low mutual coupling design for array applications," *IEEE Transactions on Antennas and Propagation*, vol. 51, no. 10, part 2, pp. 2936–2946, 2003.

22 F. Yang and Y. Rahmat-Samii, "A low profile circularly polarized curl antenna over electromagnetic band-gap (EBG) surface," *Microwave and Optical Technology Letters*, vol. 31, no. 4, pp. 264–267, 2001.

23 A. R. Weily, L. Horvath, K. P. Esselle, B. C. Sanders, and T. S. Bird, "A planar resonator antenna based on a woodpile EBG material," *IEEE Transactions on Antennas and Propagation*, vol. 53, no. 1, pp. 216–23, 2005.

24 P. Salonen, M. Keskilammi, and L. Sydanheimo, "A low-cost 2.45 GHz photonic band-gap patch antenna for wearable systems," *Proceedings of the 11th International Conference on Antennas and Propagation ICAP 2001*, pp. 719–724, 17–20 April 2001, Manchester, UK.

25 R. F. Jimenez Broas, D. F. Sievenpiper, and E. Yablonovitch, "A high-impedance ground plane applied to cellphone handset geometry," *IEEE Transactions on Microwave Theory and Techniques*, vol. 49, no. 7, pp. 1262–1265, 2002.

26 J. I. Park, C. S. Kim, J. Kim, J. S. Park, Y. Qian, D. Ahn, and T. Itoh, "Modeling of a photonic bandgap and its application for the low pass filter design," *Proceedings Asia Pacific Microwave Conference*, pp. 331–334, 1999, Singapore.

27 C. S. Kim, J. S. Park, D. Ahn, and J. B. Lim, "A novel 1-D periodic defected ground structure for planar circuits," *IEEE Microwave and Wireless Components Letters,* vol. 10, no. 4, pp. 131–133, 2000.

28 A. B. Abdel-Rahman, A. K. Verma, A. Boutejdar, and A. S. Omar, "Control of bandstop response of hi-lo microstrip low-pass filter using slot in ground plane," *IEEE Transactions on Microwave Theory and Techniques,* vol. 52, no. 3, pp. 1008–1013, 2004.

29 C. S. Kim, J. S. Lim, S. Nam , K. Y. Kang, and D. Ahn, "Equivalent circuit modelling of spiral defected ground structure for microstrip line," *Electronics Letters,* vol. 38, no. 19, pp. 1109–1110, 2002.

30 D. J. Woo, T. K. Lee, J. W. Lee, C. S. Pyo, and W. K. Choi, "Novel U-slot and V-slot DGSs for bandstop filter with improved Q factor," *IEEE Transactions on Microwave Theory and Techniques,* vol. 54, no. 6, pp. 2840–2847, 2006.

31 M. K. Mandal and S. Sanyal, "A novel defected ground structure for planar circuits," *IEEE Microwave and Wireless Components Letters,* vol. 16, no. 2, pp. 93–95, 2006.

32 X. Wang, B. Wang, H. Zhang, and K. J. Chen, "A tunable bandstop resonator based on a compact slotted ground structure," *IEEE Transactions on Microwave Theory and Techniques,* vol. 55, no. 9, pp. 1912–1917, 2007.

33 S. Y. Huang and Y. H. Lee, "A compact E-shaped patterned ground structure and its applications to tunable bandstop resonator," *IEEE Transactions on Microwave Theory and Techniques,* vol. 57, no. 3, pp. 657–666, 2009.

34 H. J. Chen, T. H. Huang, C. S. Chang, L. S. Chen, N. F. Wang, Y. H. Wang, and M. P. Houng, "A novel cross-shape DGS applied to design ultra-wide stopband low-pass filters," *IEEE Microwave and Wireless Components Letters,* vol. 16, no. 5, pp. 252–254, 2006.

35 D. Guha, S. Biswas, M. Biswas, J. Y. Siddiqui, and Y. M. M. Antar, "Concentric ring-shaped defected ground structures for microstrip applications," *IEEE Antennas and Wireless Propagation Letters,* vol. 5, pp. 402–405, 2006.

36 A. K. Arya, A. Patnaik, and M. V. Kartikeyan, "Microstrip patch antenna with skew-F shaped DGS for dual band operation," *Progress In Electromagnetics Research M,* vol. 19, pp. 147–160, 2011.

37 S. N. Burokur, M. Latrach, and S. Toutain, "A novel type of microstrip coupler utilizing a slot split ring resonators defected ground plane," *Microwave and Optical Technology Letters,* vol. 48, no. 1, pp. 138–141, 2006.

38 Z. Z. Hou, "Novel wideband filter with a transmission zero based on split-ring resonator DGS," *Microwave and Optical Technology Letters,* vol. 50, no. 6, pp. 1691–1693, 2008.

39 H. W. Liu, Z. F. Li, and X. W. Sun, "A novel fractal defected ground structure and its application to the low-pass filter," *Microwave and Optical Technology Letters,* vol. 39, no. 6, pp. 453–456, 2003.

40 M. Kufa and Z. Raida, "Lowpass filter with reduced fractal defected ground structure," *Electronics Letters*, vol. 49, no. 3, pp. 199–201, 2013.

41 H.-X. Xu, G.-M. Wang, C.-X. Zhang, and Q. Peng, "Hilbert-shaped complementary single split ring resonator and low-pass filter with ultra-wide stopband, excellent selectivity and low insertion-loss," *International Journal of Electronics Communications (AEÜ)*, vol. 65, no. 11, pp. 901–905, 2011.

42 "High Frequency Structure Simulator (HFSS) v.15.0," *Ansys*.

43 D. Ahn, J. S. Park, C. S. Kim, J. Kim, Y. Qian, and T. Itoh, "A design of the low-pass filter using the novel microstrip defected ground structure," *IEEE Transactions on Microwave Theory and Techniques*, vol. 49, no. 1, pp. 86–93, 2001.

44 A. Boutejdar, A. Omar, A. Batmanov, and E. Burte, "Design of Compact Low-Pass Filter Using Cascaded Arrowhead-DGS and Multilayer-Technique," *2009 German Microwave Conference*, pp. 1–4, 2009, Macau, China.

45 S. Dwari and S. Sanyal, "Size reduction and harmonic suppression of microstrip branch-line coupler using defected ground structure," *Microwave and Optical Technology Letters* 48, no. 10, pp. 1966–1969, 2006.

46 A. Boutejdar, M. Makkey, A. Elsherbini, and A. Omar, "Design of compact stop-band extended microstrip low-pass filters by employing mutual coupled square-shaped defected ground structures," *Microwave and Optical Technology Letters*, vol. 50, no. 4, pp. 1107–1111, 2008.

47 A. M. E. Safwat, F. Podevin, P. Ferrari, and A. Vilcot, "Tunable bandstop defected ground structure resonator using reconfigurable dumbbell-shaped coplanar waveguide," *IEEE Transactions on Microwave Theory and Techniques*, vol. 54, no. 9, pp. 3559–3564, 2006.

48 J. Wang, H. Ning, and L. Mao, "A compact reconfigurable bandstop resonator using defected ground structure on coplanar waveguide," *IEEE Antennas and Wireless Propagation Letters*, vol. 11, pp. 457–459, 2012.

49 A. Boutejdar, A. Sherbini, W. Ali, S. Fouad, L. Ahmed, A. Omar, "Design of compact microstrip lowpass filters using coupled half circle defected ground structures (DGSs)," *Proceedings of Antennas and Propagation Society International Symposium, APS-2008*, 1–4 July 2008, San Diego, CA.

50 S.-W. Ting, K.-W. Tam, and R. P. Martins, "Compact microstrip quasi-elliptic bandpass filter using open-loop dumbbell shaped defected ground structure," *IEEE MTT-S International Microwave Symposium Digest*, pp. 527–530, 2006, San Francisco, CA, USA.

51 F. Zhang, J. Gu, L. Shi, C. Li, and X. Sun, "Design of UWB lowpass filter using a novel defected ground structure," *Microwave and Optical Technology Letters* vol. 48, no. 9, pp. 1805–180, 2006.

52 A. Boutejdar, A. Elsherbini, and A. Ornar, "Design of a novel ultra-wide stopband lowpass filter using H-defected ground structure," *Microwave and Optical Technology Letters*, vol. 50, no. 3, pp. 771–775, 2008.

53 M. M. Bait-Suwailam, O. F. Siddiqui, and O. M. Ramahi, "Mutual coupling reduction between microstrip patch antennas using slotted-complementary split-ring resonators," *IEEE Antennas and Wireless Propagation Letters*, vol. 9, pp. 876–878, 2010.

54 M. K. Mandal and S. Sanyal, "U-shaped microstrip structure to decrease DGS resonance frequency," *36th European Microwave Conference*, 2006, Manchester, UK.

55 H. Liu, Z. Li, and X. Sun, "Compact defected ground structure in microstrip technology," *Electronics Letters*, vol. 41, no. 3, pp. 132–134, 2005.

56 J. S. Lim, C.S. Kim, Y.T. Lee, D. Ahn, and S. Nam, "Design of lowpass filters using defected ground structure and compensated microstrip line," *Electronics Letters*, vol. 38, no. 22, pp. 1357–1358, 2002.

57 J.-U. Kim, K.-S. Kim, S.-J. Lee, J.-S. Lim, D. Ahn, K.-H. Park, and K.-S. Kim, "A new defected ground structure with islands and equivalent circuit model," *2005 Asia-Pacific Microwave Conference (APMC) Proceedings*, vol. 1, 4–7 December 2005, Suzhou.

58 K. Song, Y.-Z. Yin, X. Yang, J.-Y. Deng, and H.-H. Xie, "Compact LPF with pair of coupling slots for wide stopband suppression," *Electronics Letters*, vol. 46, no. 13, pp. 922–924, 2010.

59 Y. Jeong, J. S. Lim, and D. Ahn, "Design of novel attenuator structure with quad spiral shaped defected ground structure," *IEEE MTT-S International Microwave Symposium Digest*, pp. 1231–1234, 2005, Long Beach, CA, USA.

60 J. S. Lim, C. S. Kim, Y. K. Lee, D. Ahn, and S. Nam, "A spiral-shaped defected ground plane structure for coplanar waveguide," *IEEE Microwave and Wireless Components Letters*, vol. 12, no. 9, pp. 312–330, 2002.

61 Y. Sung, M. Kim, and Y. Kim, "Harmonics reduction with defected ground structure for a microstrip patch antenna," *IEEE Antennas and Wireless Propagation Letters*, vol. 2, no. 8, pp. 111–113, 2003.

62 S.-W. Ting, K.-W. Tam, and R. P. Martins, "Miniaturized microstrip lowpass filter with wide stopband using double equilateral U-shaped defected ground structure," *IEEE Microwave and Wireless Components Letters*, vol. 16, no. 5, pp. 240–242, 2006.

63 D. Guha, S. Biswas, T. Joseph, and M. T. Sebastian, "Defected ground structure to reduce mutual coupling between cylindrical dielectric resonator antennas," *Electronics Letters*, vol. 44, no. 1, pp. 836–837, 2008.

64 C. Kumar and D. Guha, "Nature of cross-polarized radiation from probe fed circular microstrip antenna and their suppression using different geometries

of DGS," *IEEE Transactions on Antennas and Propagation*, vol. 60, no. 1, pp. 92–101, 2012.

65 D. Guha, C. Kumar, and S. Pal, "Improved cross-polarization characteristics of circular microstrip antenna employing arc-shaped defected ground structure (DGS)," *IEEE Antennas and Wireless Propagation Letters*, vol. 08, pp. 1367–1369, 2009.

66 S. Biswas and D. Guha, "Isolated open-ring defected ground structure to reduce mutual coupling between circular microstrips: characterization and experimental verification," *Progress In Electromagnetics Research M*, vol. 29, pp. 109–119, 2013.

67 S. Biswas and D. Guha, "Stop-band characterization of an isolated DGS for reducing mutual coupling between adjacent antenna elements and experimental verification for dielectric resonator antenna array," *International Journal of Electronics and Communications (AEÜ)*, vol. 65, no. 4, pp. 319–322, 2011.

68 S. Biswas, D. Guha, and C. Kumar, "Control of higher harmonics and their radiations in microstrip antennas using compact defected ground structures," *IEEE Transactions on Antennas and Propagation*, vol. 61, no. 6, pp. 3349–3353, 2013.

69 A. Balalem, A. R. Ali, J. Machac, and A. Omar, "Quasi-elliptic microstrip low-pass filters using an interdigital DGS slot," *IEEE Microwave and Wireless Components Letters*, vol. 17, no. 8, pp. 586–588, 2007.

70 J. X. Chen, J.L. Li, K.C. Wan, and Q. Xue, "Compact quasi-elliptic function filter based on defected ground structure," *IEE Proceedings – Microwave, Antennas and Propagation*, vol. 153, no. 4, pp. 320–324, 2006.

71 E. K. I. Hamad, A. M. E. Safwat, and A. Omar, "L-shaped defected ground structure for coplanar waveguide," *IEEE International Symposium on Antennas and Propagation and USNC-URSI Radio Science Meeting (APS/URSI)*, pp. 3–8, 2005, Washington, DC.

72 S. Lee, S. Oh, W. S. Yoon, et al., "A CPW bandstop filter using double hairpin-shaped defected ground structures with a high Q factor," *Microwave and Optical Technology Letters*, vol. 58, no. 6, pp. 1265–1268, 2016.

73 D. Guha, M. Biswas, and Y. M. M. Antar, "Microstrip patch antenna with defected ground structure for cross polarization suppression," *IEEE Antennas and Wireless Propagation Letters*, vol. 4, pp. 455–458, 2005.

74 A. Ghosh, D. Ghosh, S. Chattopadhyay, and L. L. K. Singh, "Rectangular microstrip antenna on slot-type defected ground for reduced cross-polarized radiation," *IEEE Antennas and Wireless Propagation Letters*, vol. 14, pp. 321–324, 2015.

75 D. Guha, C. Kumar, and S. Pal, "Improved cross-polarization characteristics of circular microstrip antenna employing arc-shaped defected ground structure (DGS)," *IEEE Antennas and Wireless Propagation Letters*, vol. 8, pp. 1367–1369, 2009.

76 C. Kumar and D. Guha, "Defected ground structure (DGS)-integrated rectangular microstrip patch for improved polarisation purity with wide impedance bandwidth," *IET Microwaves, Antennas and Propagation*, vol. 8, no. 8, pp. 589–596, 2014.

77 F. Y. Zulkili, E. T. Rahardjo, and D. Hartanto, "Radiation properties enhancement of triangular patch microstrip antenna array using hexagonal defected ground structure," *Progress In Electromagnetics Research M*, vol. 5, pp. 101–109, 2008.

78 H. W. Liu, Z. F. Li, X. W. Sun, and J. F. Mao, "An improved 1-D periodic defected ground structure for microstrip line," *IEEE Microwave and Wireless Components Letters*, vol. 14, no. 4, pp. 180–182, 2004.

79 J. S. Lim, Y. T. Lee, C. S. Kim, D. Ahn, and S. Nam, "A vertically periodic defected ground structure and its application in reducing the size of microwave circuits," *IEEE Microwave and Wireless Components Letters*, vol. 12, no. 12, pp. 479–481, 2002.

80 D. J. Woo and T. K. Lee, "Suppression of harmonics in Wilkinson power divider using dual-band rejection by asymmetric DGS," *IEEE Transactions on Microwave Theory and Techniques*, vol. 53, no. 6, pp. 2139–2144, 2005.

81 S. K. Parui and S. Das, "An asymmetric defected ground structure with elliptical response and its application as a lowpass filter," *International Journal of Electronics and Communications (AEU.)*, vol. 63, no. 6, pp. 483–49, 2009.

82 T. Moyra, K. Parui, and S. Das, "Application of a defected ground structure and alternative transmission line for designing a quasi-elliptic lowpass filter and reduction of insertion loss," *International Journal of RF and Microwave Computer-Aided Engineering*, vol. 20, no. 6, pp. 682–688, 2010.

83 D. Guha, S. Biswas, and Y. M. M. Antar, "Defected ground structure for microstrip antennas," *Microstrip and Printed Antennas: New Trends, Techniques and Applications*, Eds. D. Guha and Y. M. M. Antar, *John Wiley & Sons*, United Kingdom, pp. 387–434, 2011.

84 D. Guha, S. Biswas, and C. Kumar, "Printed antenna designs using defected ground structures: a review of fundamentals and state-of-the-art developments," *Forum for Electromagnetic Research Methods and Application Technologies (FERMAT)*, 2014.

85 J.-S. Lim, H.-S. Kim, J.-S. Park, D. Ahn, and S. Nam, "A power amplifier with efficiency improved using defected ground structure," *IEEE Microwave and Wireless Components Letters*, vol. 11, no. 4, pp. 170–172, 2001.

86 Y. C. Jeong, S. G. Jeong, J. S. Lim, and S. Nam, "A new method to suppress harmonics using $\lambda/4$ bias line combined by defected ground structure in

power amplifiers," *IEEE Microwave and Wireless Components Letters*, vol. 13, no. 12, pp. 538–540, 2003.

87 J. S. Lim, J. S. Park, Y. T. Lee, D. Ahn, and S. Nam, "A application of defected ground structure in reducing the size of amplifiers," *IEEE Microwave and Wireless Components Letters*, vol. 12, no. 7, pp. 261–263, 2002.

88 H-J. Choi, J-S. Lim, and Y-C. Jeong, "A new design of Doherty amplifiers using defected ground structure," *IEEE Microwave and Wireless Components Letters*, vol. 16, no. 12, pp. 687–689, 2006.

89 Y.-T. Lee, J.-S. Lim, J.-S. Park, D. Ahn, and S. Nam, "A novel phase noise reduction technique in oscillators using defected ground structure," *IEEE Microwave and Wireless Components Letters*, vol. 12, no. 2, pp. 39–41, 2002.

90 J.-S. Park and M.-S. Jung, "A novel defected ground structure for an active device mounting and its application to a microwave oscillator," *IEEE Microwave and Wireless Components Letters*, vol. 14, no. 5, pp. 198–200, 2004.

91 J. Jinse, D. J. Woo, C. Choon-Sik, and L. Taek-Kyung, "Power enhancement of microwave oscillator using a high-Q spiral-shaped DGS resonator," *Asia-Pacific Microwave Conference (APMC)*, pp. 635–640, 2006, Yokohama, Japan.

92 K.-B. Kim, T.-S. Yun, and J.-C. Lee, "A new single balanced diode mixer with DGS cell lowpass filter," *Microwave and Optical Technology Letters*, vol. 43, no. 5, pp. 423–425, 2005.

93 C. W. Ryu, C. S. Cho, J. W. Lee, and J. Kim, "New balanced self oscillating mixer using DGS resonator," *Proceedings of the 37th European Microwave Conference (EuMA)*, pp. 648–651, October 2007, Germany.

94 Y. J. Sung, C. S. Ahn, and Y.-S. Kim, "Size reduction and harmonic suppression of rat-race hybrid coupler using defected ground structure," *IEEE Microwave and Wireless Components Letters,* vol. 14, no. 1, pp. 7–9, 2004.

95 J. S. Lim, C. S. Kim, J. S. Park, D. Ahn, and S. Nam, "Design of 10 dB 90° branch line coupler using microstrip line with defected ground structure," *Electronics Letters*, vol. 36, no. 21, pp. 1784–1785, 2000.

96 J. S. Lim, S. W. Lee, C. S. Kim, J. S. Park, D. Ahn, and S. Nam, "A 4: 1 unequal Wilkinson power divider," *IEEE Microwave and Wireless Components Letters*, vol. 11, no. 3, pp. 124–126, 2001.

97 Y.-C. Jeong and J.-S. Lim, "A novel frequency Doubler using feedforward technique and defected ground structure," *IEEE Microwave and Wireless Components Letters*, vol. 14, no. 12, pp. 557–559, 2004.

98 H. Liu, T. Yoshimasu, S. Kurachi, J. Chen, Z. Li, and X. Sun, "A novel microstrip diplexer design using defected ground structure," *Proceedings International Conference on Communication, Circuits & Systems*, vol. 2, pp. 1099–1100, 2005, Hong Kong, China.

99 S. M. Han, C.-S. Kim, D. Ahn, and T. Itoh, "Phase shifter with high phase shifts using defected ground structures," *Electronics Letters*, vol. 41, no. 04, 2005.

100 S. Hekal, A. B. Abdel-Rahman, H. Jia, A. Allam, A. Barakat, T. Kaho, and R. K. Pokharel, "Compact wireless power transfer system using defected ground bandstop filters," *IEEE Microwave and Wireless Components Letters*, vol. 26, no. 10, pp. 849–851, 2016, doi: https://doi.org/10.1109/LMWC.2016.2601300.

101 S. Hekal, A. B. Abdel-Rahman, H. Jia, A. Allam, A. Barakat, and R. K. Pokharel, "A novel technique for compact size wireless power transfer applications using defected ground structures," *IEEE Transactions on Microwave Theory and Techniques*, vol. 65, no. 2, pp. 591–599, 2017.

2

Theoretical Analysis and Modeling

2.1 Introduction

The evolution of DGS along with the physical insight into the working principle has been discussed in the previous chapter. It is not difficult to realize the overwhelming popularity of this technology that developed over the last two decades spreading its prolific applications from printed circuits to planar antennas. However, only the glimpses of some elementary DGS shapes were discussed in Chapter 1.

In the initial phase of development, the commercial solvers were used to examine the transmission characteristics of a printed transmission line bearing a DGS underneath. Thus, a trial-and-error method was the only option to optimize the geometry and hence the engineers seriously felt the need of accurate modeling techniques along with suitable design equations. The first attempt dates to 2001 [1] which proposed an equivalent LC circuit to analyze a DGS. A qualitative understanding of the same has been discussed in Chapter 1. The equivalent circuit representations indeed ease out the design procedure especially when more than one DGS unit is involved. Each DGS unit may now be replaced by its equivalent circuit and analyzed using a circuit simulator which is computationally inexpensive and fast. A brief account of DGS modeling techniques along with the parameter extraction methods has been presented in this chapter.

2.2 LC and RLC Modeling

A simple dumbbell-shaped DGS beneath a $50\,\Omega$ microstrip line and its transmission characteristics as the function of DGS parameters are shown in Figure 2.1. An attenuation pole along with a cutoff embodies the typical signature of resonance in a parallel LC circuit. This indeed gives rise to the idea of the equivalent L and C combination to model a DGS as depicted in Figure 2.2.

Defected Ground Structure (DGS) Based Antennas: Design Physics, Engineering, and Applications,
First Edition. Debatosh Guha, Chandrakanta Kumar, and Sujoy Biswas.

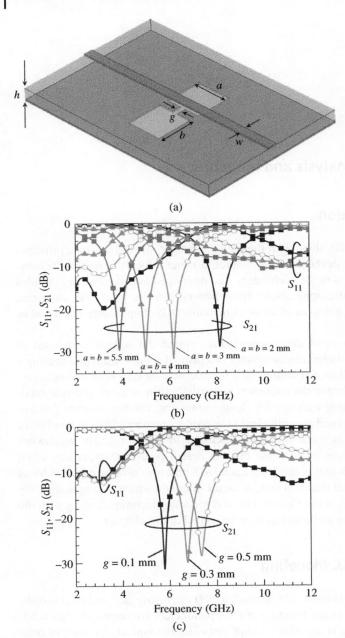

Figure 2.1 A 50 Ω microstrip line integrated with a dumbbell-shaped DGS and its S-parameters [1]: (a) An isometric view, substrate thickness $h = 1.575$ mm with $\varepsilon_r = 9.8$, and line width $w = 1.57$ mm; (b) Simulated S-parameters with $g = 0.15$ mm and varying a, b; (c) Simulated S-parameters with $a = b = 2.5$ mm and varying g. Source: Adapted from Ahn et al. [1].

Figure 2.2 LC equivalent circuit of a single cell dumbbell-shaped DGS. Source: Adapted from Ahn et al. [1].

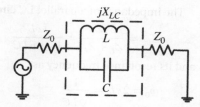

As discussed in Chapter 1, this equivalent inductance and capacitance are actually contributed caused by two different sections of a dumbbell DGS. The conduction current surrounding the square heads produces an equivalent L and the gap-coupled electric fields across the narrow connecting slot results in an equivalent C. The simulated S_{11} endorses this feature. Figure 2.1b reveals that the attenuation poles shift toward the lower frequency with the increase in square head dimension, i.e. due to the increase in equivalent L that lowers the resonance frequency. On the other hand, Figure 2.1c shows shifting of the attenuation poles towards right with gradual increase the gap dimension. An increase in gap reduces its equivalent capacitance C which pushes the resonance toward the right.

2.2.1 Equivalent Circuit Parameter Extraction

A functional circuit representation for a DGS needs extraction of equivalent L and C values for a given geometry. The parameter extraction technique has been demonstrated here for a typical dumbbell-shaped DGS etched on 31 mil RT Duroid 5880 substrate with 2.2 relative permittivity and deployed beneath a 50 Ω microstrip line [1]. The simulated transmission characteristics for a set of DGS parameters are shown in Figure 2.3, where the attenuation pole appears near 8 GHz along with a cut-off frequency f_c at 3.87 GHz.

Figure 2.3 Simulated S-parameters versus frequency of the 50 Ω microstrip line integrated with a dumbbell-shaped DGS with $a = b = 5$ mm and $g = 0.5$ mm. Substrate thickness $h = 31$ mil and dielectric constant $\varepsilon_r = 2.2$. Source: Adapted from Ahn et al. [1].

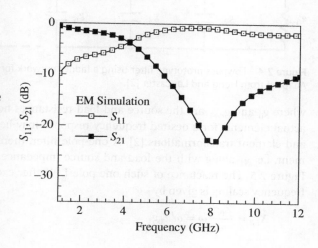

The impedance of a parallel LC circuit is expressed as

$$Z_{LC} = \frac{Z_L Z_C}{Z_L + Z_C} = \frac{j\omega L}{1 - \omega^2 LC} \tag{2.1}$$

and its resonance frequency as

$$\omega_0 = 2\pi f_0 = \frac{1}{\sqrt{LC}} \tag{2.2}$$

The reactance of this circuit is

$$X_{LC} = \frac{1}{\omega_0 C \left(\frac{\omega_0}{\omega} - \frac{\omega}{\omega_0} \right)} \tag{2.3}$$

If this LC circuit is meant for representing a DGS with its response as in Figure 2.3, it should behave like a low-pass filter (LPF) at the cutoff and a band stop filter at resonance. Therefore, to extract the values of L and C, both behaviors need to be taken into account. At first, the reactance of the equivalent circuit is equated to the reactance of a one-pole Butterworth type LPF at the 3-dB cut-off frequency. Then the attenuation pole is equated to the resonance frequency ω_0.

To deal with a one-pole Butterworth type LPF, one must start from an n-pole low-pass prototype filter. Such a prototype is shown in Figure 2.4 [2] which comprises normalized elements g_1 to g_n with the angular cutoff frequency set to unity. The g_i values are given by

$$g_i = 2\sin\left(\frac{(2i-1)\pi}{2n}\right) \qquad i = 1 \text{ to } n \tag{2.4}$$

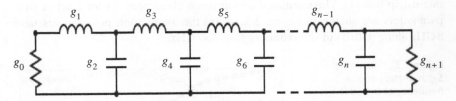

Figure 2.4 Lowpass prototype filter using a ladder network for even *n* values. Source: Adapted from Hong and Lancaster [2].

where g_0 and g_{n+1} are the source and load resistances normalized to unity. The actual elements for a desired frequency response may be obtained by frequency and element transformations [2]. A one-pole filter, therefore, contains one element, i.e. g_1 along with the load and source impedance g_0 and g_2 as shown in Figure 2.5. The reactance of such one pole LPF after executing impedance and frequency scaling is given by

$$X_{LB} = \frac{Z_0 g_1}{\omega_c} \omega = \omega' Z_0 g_1 \tag{2.5}$$

where Z_0 is the impedance scaling for making the source and load impedances unity, ω_c is the cutoff angular frequency and ω' is the normalized angular frequency.

Figure 2.5 One-pole Butterworth prototype Lowpass filter. Source: Adapted from Ahn et al. [1].

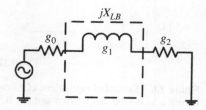

To match the response of one-pole LPF with that of the equivalent circuit of our interest, their reactance at the cutoff should follow

$$X_{LC}|_{\omega=\omega_c} = X_{LB}|_{\omega'=1} \tag{2.6}$$

resulting in

$$\frac{1}{\omega_0 C\left(\frac{\omega_0}{\omega_c} - \frac{\omega_c}{\omega_0}\right)} = Z_0 g_1 \tag{2.7}$$

or,

$$C = \frac{\omega_c}{Z_0 g_1 (\omega_0^2 - \omega_c^2)} \tag{2.8}$$

Equation (2.2) with the help of (2.8) yields

$$L = \frac{1}{4\pi^2 f_0^2 C} \tag{2.9}$$

Now with $g_1 = 2$ for one-pole LPF and $Z_0 = 50\,\Omega$, the equivalent parameters are

$$C = \frac{5f_c}{\pi\left(f_0^2 - f_c^2\right)} \text{ pF} \tag{2.10}$$

$$L = \frac{250}{(\pi f_0)^2 C} \text{ nH} \tag{2.11}$$

where f_0 and f_c are the attenuation pole and cut-off frequency respectively in GHz. The transmission characteristics of a concerned DGS thus help in finding out f_0 and f_c and hence the equivalent L and C values. The example shown in Figure 2.3 indicates $f_0 = 8$ GHz and $f_c = 3.87$ GHz which with the help of (2.10) and (2.11) extracts equivalent L and C as 3.139 nH and 0.126 pF, respectively. The resulting equivalent circuit is shown in Figure 2.6 and its S-parameter response has been compared with EM simulated values in Figure 2.7. An excellent mutual agreement is revealed. This technique has been applied to a wide range of DGS dimensions and resulting values have been furnished in Table 2.1.

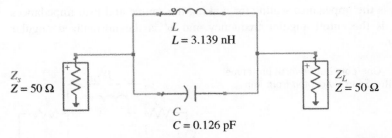

Figure 2.6 Extracted equivalent LC circuit for the dumbbell-shaped DGS shown in Figure 2.1a with $a = b = 5$ mm and $g = 0.5$ mm.

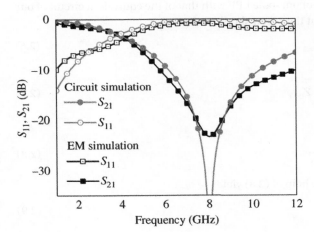

Figure 2.7 Simulated S-parameters for the extracted LC equivalent circuit shown in Figure 2.6 compared with those obtained for a DGS integrated 50 Ω microstrip line as in Figure 2.1(a).

Table 2.1 Circuit elements and characteristics for the equivalent circuit in Figure 2.2 for different dimensions of the dumbbell-shaped DGS.

Equivalent circuit elements and characteristics	Dumbbell-shaped DGS: Dimensions in mm					
	$g = 0.15$			$a = b = 2.5$		
	$a = b = 2$	$a = b = 3$	$a = b = 4$	$g = 0.1$	$g = 0.3$	$g = 0.5$
Inductance (nH)	0.7056	1.2940	1.991	1.273	1.345	1.348
Capacitance (pF)	0.5409	0.513	0.5155	0.5978	0.4124	0.3499
Cut-off frequency f_c (GHz)	6.81	4.82	3.66	4.59	5.1	5.4
Attenuation pole f_0 (GHz)	8.15	6.18	4.97	5.77	6.76	7.33

2.2.2 Utilization of the Extracted LC for n-Pole DGS Filter Design

The equivalent circuit extracted in Section 2.2.1 may now be treated as a "unit" for realizing a more complex n-pole DGS filter using only the circuit simulations and without executing any EM-based full wave analysis. A maximally flat three-pole LPF with its specified cutoff at 2.3 GHz has been demonstrated as a representative example. A indicated earlier, its prototype consists of three elements g_1, g_2, and g_3 along with the source and load impedances $g_0 = g_4 = 1$. Figure 2.8a shows the lumped circuit representation based on the impedance and frequency scaling as discussed earlier. A DGS representing an equivalent parallel LC circuit acts as an inductor below resonance which may be expressed as

$$L|_{f<f_0} = \frac{L_{eq}}{\left[1 - \left(\frac{f}{f_0}\right)^2\right]} \text{ nH} \tag{2.12}$$

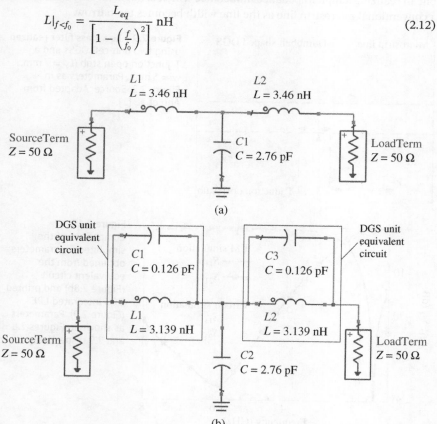

(a)

(b)

Figure 2.8 (a) Schematic diagram of a 3-pole Butterworth lowpass filter using lumped circuit elements with cutoff frequency of 2.3 GHz. (b) Equivalent circuit of a dumbbell DGS is employed to modify the circuit in (a). Cut-off frequency 2.3 GHz and ripple 0.01 dB. Source: Adapted from Ahn et al. [1].

This inductance is a slowly varying function of frequency below cutoff and therefore, L_1 and L_2 in Figure 2.8a can be substituted by an equivalent DGS. Figure 2.8b clarifies the scenario. It actually represents a pair of DGS units in series as shown in Figure 2.9. Lumped capacitor C_2 of Figure 2.8b is realized by introducing an open-ended stub (Figure 2.9). Related calculation uses Richards Transforms and Kuroda identities [2]. Figure 2.10 compares the scattering parameters of the LPF circuit (Figure 2.8b) obtained by circuit simulation with those of the equivalent DGS configuration in Figure 2.9. They reveal an excellent mutual agreement. The LPF exhibits more than 20 dB rejection up to 8 GHz. This DGS filters exhibit deeper and wider stopband compared to a conventional LPF. The DGS filter also allows one in realizing a high impedance inductance which is not at all easy to obtain by a conventional microstrip line as the line width becomes too narrow.

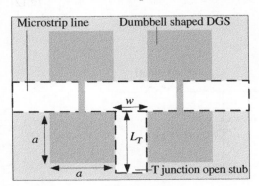

Figure 2.9 Low-pass filter realized using two unit cell DGSs and a T-junction open stub ($L_T = 7$ mm, $w = 5$ mm). Parameters as in Figure 2.2. Source: Adapted from Ahn et al. [1].

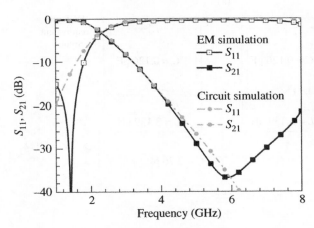

Figure 2.10 Comparison of the simulated S-parameters obtained from the equivalent circuit (Figure 2.8b) and printed DGS integrated LPF (Figure 2.9). Parameters as shown in Figures 2.8 and 2.9 respectively.

Therefore, once the LC equivalent of a basic DGS is extracted, they can be reliably utilized in designing a higher order filter and its DGS version without any time consuming iteration or trial and error approach using EM simulation.

2.2.3 RLC Circuit Modeling

The parallel LC circuit modeling of a DGS does not account for any losses although it is quite obvious in the form of radiation, leakage, conductor/dielectric heating in an RF circuit. The comparative study in Figure 2.7 endorses the same. Thus, a realistic model should incorporate an equivalent resistance R in parallel with L and C [3] as shown in Figure 2.11a. This loss resistance R has been extracted from the transmission line equation [3] as

$$R = \frac{2Z_0}{\sqrt{\frac{1}{|S_{11}(\omega)|^2} - \left(2Z_0\left(\omega C - \frac{1}{\omega L}\right)\right)^2 - 1}} \qquad (2.13)$$

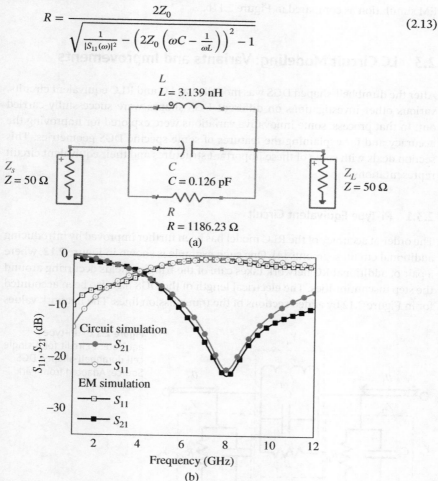

Figure 2.11 (a) Extracted equivalent RLC circuit for the dumbbell-shaped DGS shown in Figure 2.1 with $a = b = 5$ mm and $g = 0.5$ mm; (b) circuit simulated S-parameters for the above RLC circuit compared with EM Simulated S-parameters for the DGS in Figure 2.1a.

with

$$S_{11}(\omega) = \frac{Z_{in} - Z_0}{Z_{in} + Z_0} \tag{2.14}$$

For the given values of attenuation pole and cutoff frequency related to Figure 2.3, $|S_{11}(\omega)| \approx 0.9063$. The values of L and C obtained from (2.10) and (2.11) readily help in calculating $R = 1186.23\,\Omega$. The circuit simulation for the equivalent RLC circuit (Figure 2.11a) reveals much closer approximation to the EM simulation as compared in Figure 2.11b.

2.3 LC Circuit Modeling: Variants and Improvements

After the dumbbell-shaped DGS was modelled as LC and RLC equivalent circuits, various other investigations on different DGS shapes were successfully carried out. In that process, some innovative variations were explored for improving the accuracy and for explaining the features of some specific DGS geometries. This section deals with some of these important structures and their equivalent circuit representations.

2.3.1 Pi-Type Equivalent Circuit

The order of accuracy of the RLC model has been further improved by introducing additional circuit segments [4]. One such example is shown in Figure 2.12, where a pair of additional RC network takes care of the fringing fields occurring around the step discontinuities. The electrical length of the DGS has also been accounted for in Figure 2.12 by added sections of the transmission lines. The network values

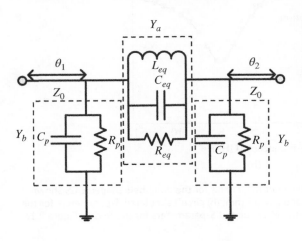

Figure 2.12 Pi-type equivalent circuit for a single cell dumbbell-shaped DGS. Source: Adapted from Park et al. [4].

are extracted by equating its ABCD parameters with the S-parameter values of the DGS characteristics [4]. The relations are given by

$$A = \frac{(1 + S_{11})(1 - S_{22}) + S_{12}S_{21}}{2S_{21}} \tag{2.15}$$

$$B = \frac{(1 + S_{11})(1 + S_{22}) - S_{12}S_{21}}{2S_{21}} \tag{2.16}$$

$$C = \frac{1}{Z_0} \frac{(1 - S_{11})(1 - S_{22}) - S_{12}S_{21}}{2S_{21}} \tag{2.17}$$

$$D = \frac{(1 - S_{11})(1 + S_{22}) + S_{12}S_{21}}{2S_{21}} \tag{2.18}$$

The ABCD parameters are commonly expressed in terms of admittance values Y_a and Y_b as indicated in Figure 2.12 for the Pi-type network as

$$A = 1 + \frac{Y_b}{Y_a} \tag{2.19}$$

$$B = \frac{1}{Y_a} \tag{2.20}$$

$$C = 2Y_b + \frac{Y_b^2}{Y_a} \tag{2.21}$$

$$D = 1 + \frac{Y_b}{Y_a} \tag{2.22}$$

These two sets of Eqs. (2.15)–(2.18) and (2.19)–(2.22) provide

$$Y_a = \frac{1}{B} = \frac{1}{R_{eq}} + jB_a \tag{2.23}$$

$$Y_b = \frac{A-1}{B} = \frac{D-1}{B} = \frac{1}{R_p} + jB_p \tag{2.24}$$

where, B_a is the susceptance of the LCR equivalent circuit corresponding to Y_a, and B_p is the susceptance of the CR circuit corresponding to Y_b.

Equation (2.23) with the help of (2.2) results in the equivalent capacitance

$$C_{eq} = \frac{B_a}{\omega_0 \left(\frac{\omega_c}{\omega_0} - \frac{\omega_0}{\omega_c} \right)} \tag{2.25}$$

The equivalent inductance is obtained from the resonance condition of (2.9) as

$$L_{eq} = \frac{1}{\omega_0^2 C_{eq}} \tag{2.26}$$

Resistances R_{eq} and R_p are calculated from the real values of Y_a and Y_b, and capacitance C_p is given by

$$C_p = \frac{B_p}{\omega_c} \tag{2.27}$$

Here the equivalent circuit parameters have been extracted for the dumbbell-shaped DGS with the same dimensions and substrate specifications as in Figure 2.1. The extracted quantities for $f_c = 3.88\,\text{GHz}$ and $f_0 = 8\,\text{GHz}$ are found to be $R_{eq} = 2330\,\Omega$, $R_p = 2560\,\Omega$, $L_{eq} = 3.43\,\text{nH}$, $C_{eq} = 0.116\,\text{pF}$, and $C_p = 0.36\,\text{pF}$. Their response based on the circuit simulation shows very close agreement with the EM simulated data as furnished in Figure 2.13. It reveals a better agreement at the cutoff frequency compared to that in Figure 2.11. Also, a better resemblance is apparent beyond resonance compared to that predicted by a simple RLC circuit (Figure 2.11).

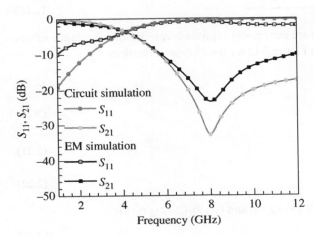

Figure 2.13 Circuit simulated S-parameters for Pi equivalent circuit in Figure 2.12 compared with the EM simulated curves for DGS integrated $50\,\Omega$ microstrip line shown in Figure 2.1a.

2.3.2 Modeling of Spiral DGS with Periodic Resonance

Spiral-shaped DGS was studied immediately after the dumbbell DGS [5]. A typical geometry is shown in Figure 2.14 where the rectangular head of a dumbbell is replaced by spiral geometry. The basic idea is to increase the equivalent inductance and capacitance. The resultant attenuation pole shifts to a much lower frequency along with a steeper rejection characteristic compared to that in a dumbbell DGS of identical size. The EM simulation-based studies in Figure 2.15 reveal multiple periodic resonances. This feature of spiral head DGS has been modelled as a short-circuited stub with characteristic impedance Z_s as shown in Figure 2.16. The overall inductance of the spiral structure is represented by L_s. These equivalent parameters can be easily extracted following the methodology discussed in Section 2.2.1. This uses one-pole Butterworth type LPF prototype shown in Figure 2.5 and the corresponding susceptance of L_B is

$$B_B = \frac{-1}{\omega L_B} \tag{2.28}$$

Figure 2.14 Schematic diagrams of spiral shape DGS beneath a microstrip transmission line. Source: Adapted from Kim et al. [5].

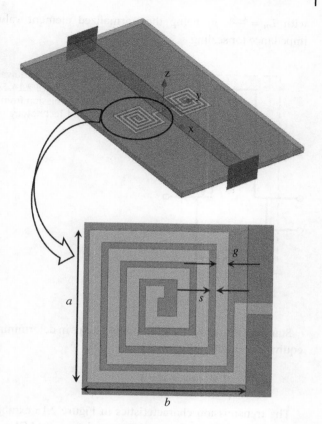

Figure 2.15 EM simulated S-parameters vs frequency for the spiral DGS integrated microstrip line as in Figure 2.14. Substrate: RT Duroid 5880 of thickness 31 mil and $\varepsilon_r = 2.2$, $a = b = 5$ mm, $g = 0.4$ mm, $S = 0.2$ mm. Source: Adapted from Kim et al. [5].

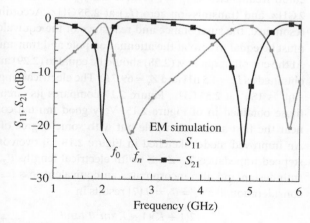

with $L_B = \frac{g_1 Z_0}{\omega_c}$, g_1 being the normalized element value, and Z_0 = the port impedance for scaling.

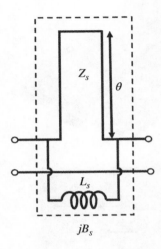

Figure 2.16 Proposed equivalent circuit for the spiral DGS depicted in Figure 2.14. Source: Adapted from Kim et al. [5] with permission from The Institution of Engineering and Technology.

Standard transmission line equation helps in determining the susceptance of the equivalent circuit in Figure 2.16 as

$$B_s = -\left(Y_s cot\theta + \frac{1}{\omega L_s}\right) \tag{2.29}$$

The transmission characteristics in Figure 2.15 exhibit lower and upper 3 dB cutoff frequencies as $f_{cl} = 1.75$ GHz and $f_{cu} = 2.14$ GHz, attenuation pole (f_0) at 2 GHz, and transmission zero (f_n) at 2.55 GHz. According to the conditions at resonance, the susceptance and reactance of the equivalent circuit (Figure 2.16) must be equal to zero at the attenuation pole and transmission zero, respectively. At the cut-off frequency, (2.28) should be equal to (2.29) and thus the solutions are obtained as $L_s = 4.5$ nH and $Z_s = 69.7\,\Omega$. The electrical length of the short-circuited stub is 180° at 2.55 GHz. Figure 2.17 compares its circuit simulation data with those obtained in of Figure 2.15. Very good mutual corroboration is observed near the first attenuation pole but with some degree of deviation beyond that. An improved model sketched in Figure 2.18 [6] overcomes this lacuna. It uses stepped impedance Z_2, Z_1, Z_2 with electrical lengths θ_2, $2\theta_1$, θ_2 for the shorted stub design and also incorporates an equivalent loss resistance R_s. A simplified consideration like $\theta_1 = \theta_2 = \theta$ [7] results in

$$Z_{in} = jZ_2 \frac{2(1 + K)(1 - Ktan^2\theta)tan\theta}{K - 2(1 + K + K^2)tan^2\theta + Ktan^4\theta} = jZ_2 F(\omega) \tag{2.30}$$

where K is the impedance ratio Z_2/Z_1. The admittance Y_S of the equivalent circuit for the spiral DGS is

$$Y_s = G_s + j\,B_s(\omega) \qquad (2.31)$$

with $G_s = 1/R_s$ and $B_s(\omega) = Y_{in} - \dfrac{j}{\omega L_s} = -jY_2\dfrac{1}{F(\omega)} - \dfrac{j}{\omega L_s}.$

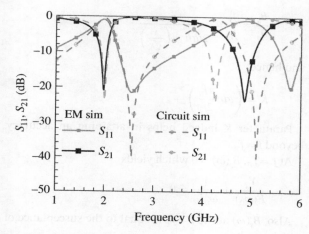

Figure 2.17 Circuit simulations for Figure 2.16 compared with the EM simulated data for the spiral DGS shown in Figure 2.14. Source: Adapted from Kim et al. [5].

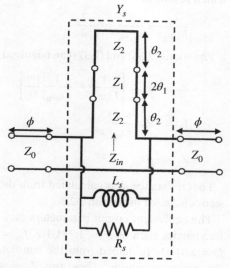

Figure 2.18 Improved equivalent circuit of the spiral DGS using stepped impedance resonator, with a special consideration $\theta_1 = \theta_2 = \theta$. Source: Adapted from Kim et al. [6].

This Y_{in} has to be infinity ($Z_{in} = 0$) for transmission zero at f_n. Setting the numerator of (2.30) equal to zero for satisfying $Z_{in} = 0$, one obtains

$$\theta_{fn} = \tan^{-1}\dfrac{1}{\sqrt{K}} \qquad (2.32)$$

At $f = f_s$, the condition $Z_{in} = 0$ is also achieved by choosing $\theta_{fs} = \pi/2$, which results in

$$tan\theta s = \infty \tag{2.33}$$

Combining (2.32) and (2.33), one can have

$$\frac{\theta_s}{\theta_n} = \frac{f_s}{f_n} = \frac{\pi}{2tan^{-1}\left(\frac{1}{\sqrt{K}}\right)} \tag{2.34}$$

and hence,

$$K = \left(tan\frac{\pi f_n}{2f_s}\right)^{-2} \tag{2.35}$$

Parameter K indeed helps in attaining an accuracy in the circuit response beyond $f = f_n$.

At $f = f_0$, $B_s(\omega) = 0$ which yields

$$\frac{Y_2}{F(\omega_0)} + \frac{1}{\omega_0 L_s} = 0 \tag{2.36}$$

Also, $B_s(\omega)$ at the $f = f_c$ is equal to the susceptance of a prototype (Figure 2.2) which results in

$$\frac{Y_2}{F(\omega_c)} + \frac{1}{\omega_c L_s} = \frac{-1}{g_1 Z_0} \tag{2.37}$$

Equations (2.36) and (2.37) can be solved to determine the following parameters:

$$Z_2 = -g_1 Z_0 \left[\frac{1}{F(\omega_c)} - \frac{1}{F(\omega_0)}\frac{\omega_0}{\omega_c}\right] \tag{2.38}$$

$$L_s = -Z_2 \frac{F(\omega_n)}{\omega_n} \tag{2.39}$$

$$Z_1 = \frac{Z_2}{K} \tag{2.40}$$

Loss resistance R_s is calculated from the simulated S-parameters following the steps discussed in Section 2.2.3.

The equivalent circuit parameters may now be extracted for a set of specified frequencies such as $f_{cl} = 1.75\,GHz$, $f_{cu} = 2.14\,GHz$, $f_0 = 2\,GHz$, $f_n = 2.55\,GHz$, $f_s = 5.8\,GHz$ (obtained from the simulated S-parameters in Figure 2.15) using the above equations. They are: $Z_1 = 48.63\,\Omega$, $Z_2 = 71\,\Omega$, electrical length $\theta = 39.568°$ at $f = f_n$, $L_s = 4.72\,nH$, $R_s = 1330\,\Omega$. Resulting circuit simulation has been compared with the EM simulated curve in Figure 2.19 revealing much improved mutual agreement over the entire frequency range compared to that observed in Figure 2.17. In the above exercise, the general assumption has been

$\theta_1 = \theta_2 = \theta = \pi/2$ at $f = f_s$. But in reality, different combinations of θ_1 and θ_2 may exist and their corresponding solutions are also possible [7].

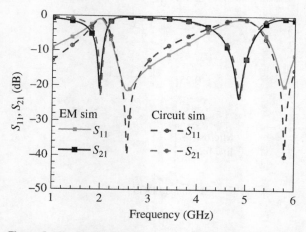

Figure 2.19 Circuit simulated S-parameters for the improved equivalent circuit in Figure 2.18 compared with the EM simulated curves for the spiral DGS shown in Figure 2.14. Parameters of the circuit: $Z_1 = 48.63\,\Omega$, $Z_2 = 71\,\Omega$, $\theta = 39.568°$ at f_n, $L_s = 4.72\,nH$, $R_s = 1330\,\Omega$. $\phi = 12°$ at f_c. DGS parameters as in Figure 2.14. Source: Adapted from Kim et al. [6].

2.3.3 Modeling of DGS with Aperiodic Stopbands

Unlike spiral DGS, there are several other shapes that exhibit multiple stopbands of aperiodic nature. The separation between the two stopbands may also be controlled by some geometrical parameters of the DGS. Two such examples on CPW and microstrip configurations are shown in Figure 2.20. For most of their applications, the frequency up to the second stopband is considered and a typical response is shown in Figure 2.21. Frequencies f_{01} and f_{02} represent the first and the second resonances with a transit frequency f_T. The interaction between two consecutive aperiodic resonances and their effect on the transmission characteristics have been modelled in Figure 2.22 [8]. A pair of parallel LC resonators account for the first and second resonances and a T-network in between is used to modeling the interaction between these two resonators.

Instead of considering a Butterworth prototype LPF, the equivalent circuit parameters have been extracted here by applying circuit theory on EM simulation-based Z-parameter data [8]:

$$C_i = \frac{1}{Z_0} \frac{1}{4\pi \Delta f_{3dB\_i}} \tag{2.41}$$

$$L_i = \frac{1}{(2\pi f_{0i})^2 C_i} \text{ where } i = 1,2 \tag{2.42}$$

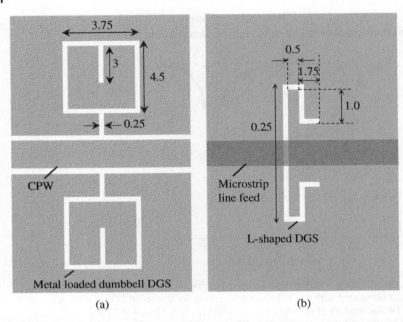

Figure 2.20 DGS geometries with multiple aperiodic stopband characteristics: (a) A 50 Ω CPW integrated with metal-loaded dumbbell DGS; (b) a 50 Ω microstrip line integrated with L-shaped DGS. All dimensions in mm, substrate thickness 1.27 mm and $\varepsilon_r = 10.8$. Source: Adapted from Hong and Karyamapudi [8].

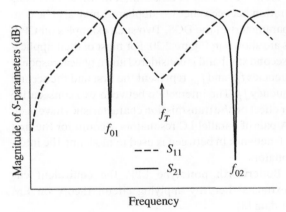

Figure 2.21 Typical response of multi-stopband DGS with two resonant frequencies f_{01} and f_{02} and the transition frequency f_T. Source: Hong and Karyamapudi [8] © [2005] IEEE.

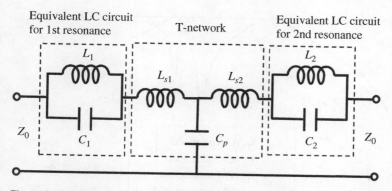

Figure 2.22 General equivalent circuit model of a DGS representing dual stopbands. Source: Hong and Karyamapudi [8] © [2005] IEEE.

$$C_p = -\frac{1}{2\pi f_T X_{21}} \tag{2.43}$$

$$L_{si} = \frac{X_{ii} - X_{21}}{2\pi f_T} + \frac{L_i}{\left(\frac{f_T}{f_{0i}}\right)^2 - 1} \quad \text{where } i = 1,2 \tag{2.44}$$

where Δf_{3dB_1} and Δf_{3dB_2} are the 3 dB bandwidth at frequency f_{01} and f_{02}, X_{11}, X_{22}, and X_{21} are the imaginary part of the Z-parameter. The 3 dB bandwidth and resonance frequencies f_{01}, f_{02} are determined from the EM simulation-based data. Equivalent Z-parameters are obtained through the standard S- to Z-parameter conversion technique [9]. Extraction of the equivalent circuit parameters for both geometries in Figure 2.20 has been executed. They use 50 Ω line on 1.27 mm thick dielectric substrates with $\varepsilon_r = 10.8$ and the extracted parameters have been furnished in Table 2.2.

Table 2.2 Extracted parameters for the metal-loaded dumbbell and L-shaped DGSs shown in Figure 2.20.

DGS Type	Extracted parameters of the DGS						
	L_1 (nH)	C_1 (pF)	L_2 (nH)	C_2 (pF)	L_{s1} (nH)	L_{s2} (nH)	C_p (pF)
Metal-loaded dumbbell	1.226	0.89966	0.2212	1.632	0.995	−0.919	0.033
L-shaped	0.4184	1.349	0.082	1.02	−0.749	0.5745	0.028

The S_{11} and S_{21} characteristics obtained using circuit simulations have been compared with the EM simulated curves in Figure 2.23 indicating excellent mutual agreements (Table 2.2).

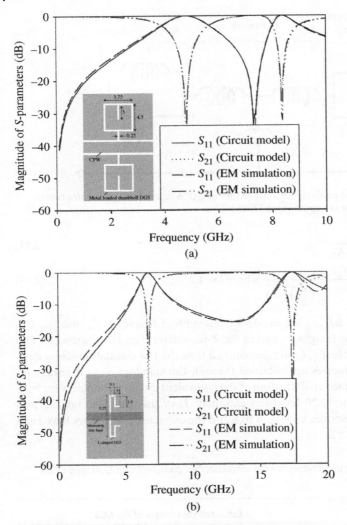

Figure 2.23 Comparison between circuit simulated and EM simulated S_{11} and S_{21} magnitude for (a) metal-loaded DGS (Figure 2.20a); (b) L-shaped DGS (Figure 2.20b). Parameters as in Figure 2.20. Source: Adapted from Hong and Karyamapudi [8].

2.3.4 Some Modifications in Modeling Approach

Different modifications over the basic LC resonator circuit model of a DGS have been explored over the time. Those are primarily required to address a few specific DGS geometries. Two such examples are shown in Figures 2.24 and 2.25.

An interdigital geometry and its equivalent circuit are depicted through Figure 2.24 [10] which introduces a series combination of L_s and C as the contributing factors of the interdigital pattern [11] in addition to L_p caused by the defect itself [10]. Figure 2.25 embodies an open square DGS and its proposed equivalent circuit [12]. The DGS is modelled as a parallel LC resonator ($L_1 C_1$) along with a series $L_2 C_2$ combination which actually takes care of the large fringing field effects in the narrow gap region. The T-network comprising L_3, L_4, L_5, and C_3 models the effect of the microstrip line on DGS. Extraction of the equivalent circuit uses the relationship between the EM simulated S-parameters and the ABCD parameters [9].

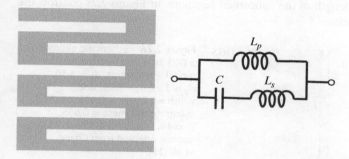

Figure 2.24 An interdigital-shaped DGS and its equivalent circuit. Source: Adapted from Balalem et al. [10].

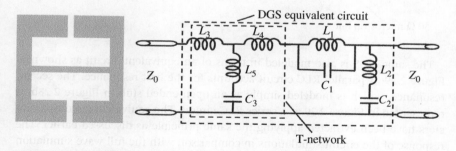

Figure 2.25 Geometry of the open square DGS and its equivalent circuit. Source: Adapted from Chen et al. [12].

2.4 Transmission Line Modeling

The transmission line model for a DGS shown was proposed in [13] and the schematic geometry is shown in Figure 2.26. The structure resonates at two different frequencies and the resulting magnetic fields around the slots are portrayed in Figure 2.27 for a useful insight into functioning of the defects. The entire slot is excited at the first resonance while only the folded arms work at the second resonance frequency. Thus, the same defect can be modeled as two different transmission lines at two resonances. The plane of symmetry marked as "1" in Figure 2.27a indicates the position of the input port for the equivalent transmission line to be considered at first resonance. Three segments have been clearly modeled and shown in Figure 2.28a. It comprises a transmission line with a pair of parallel open-ended stubs. For the 2nd resonance, the plane of symmetry is marked as "2" in Figure 2.27b and it represents the input port. Therefore, one side arm has been modeled as an open circuit transmission line as shown in the Figure 2.28b. The length of the concerned segments in Figure 2.28 controls the resonance frequencies.

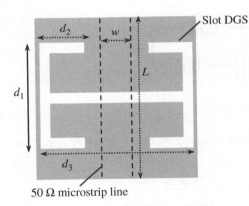

50 Ω microstrip line

Figure 2.26 Schematic geometry of a DGS analyzed by transmission line model. Parameters: $d_1 = 3.30$, $d_2 = 1.27$, $d_3 = 3.68$, slot width $= 0.254$, $w = 0.635$, $L = 37.8$, substrate thickness $= 0.635$, $\varepsilon_r = 10.2$ (all dimensions in mm). Source: Adapted from Chang et al. [13].

The same DGS is also modeled in terms of an equivalent circuit as shown in Figure 2.29. A parallel RLC circuit accounts for the first resonance. The second resonance, which is modeled simply as an open-ended stub in Figure 2.28b, is represented in Figure 2.29 as a series RLC circuit. The equivalent circuit parameters have been extracted applying the same principle as discussed earlier. The response of the circuit simulations in comparison with the full wave simulation data is shown in Figure 2.30.

Another transmission line model [14] accounts for the coupling mechanism between a DGS and a microstrip line. The geometry is shown in Figure 2.31a [14] which resonates at

$$f_m \cong m \frac{c_0}{2d\sqrt{\varepsilon_{eff}^{slot}}} \quad \text{where } m = 1, 2, 3 \ldots \quad (2.45)$$

As usual, the RF energy gets trapped in the DGS over the stopbands. The proposed equivalent circuit, shown in Figure 2.31b, is a bit different. It considers

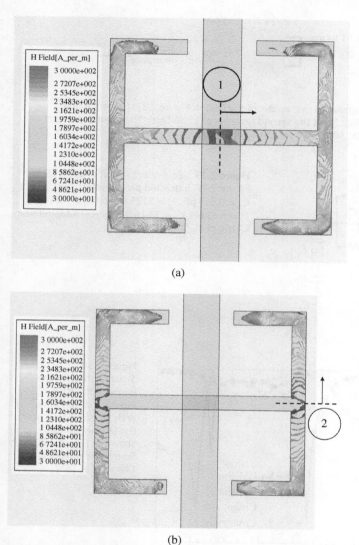

(a)

(b)

Figure 2.27 Simulated magnetic field distributions around the defects of the geometry shown in Figure 2.26: (a) obtained at the first resonance frequency of 7.19 GHz and "1" indicates the input port for transmission line model; (b) obtained at the second resonance frequency of 12.86 GHz and "2" indicates the input port for transmission line model. Parameters as shown in Figure 2.26. Source: Adapted from Chang et al. [13].

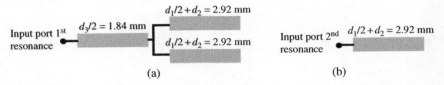

Figure 2.28 Transmission line model of the DGS in Figure 2.27: (a) at the first resonance (7.19 GHz); (b) at the second resonance (12.86 GHz). The input ports refer to Figure 2.27. Source: Adapted from Chang et al. [13].

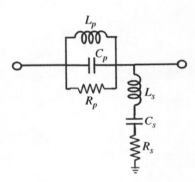

Figure 2.29 Equivalent circuit of the DGS in Figure 2.27. Extracted parameters: $L_p = 1.55$ nH, $C_p = 0.315$ pF, $R_p = 3125\,\Omega$, and $R_s = 4.41\,\Omega$, $L_s = 6.61$ nH, $C_s = 23.16$ fF. Source: Adapted from Chang et al. [13].

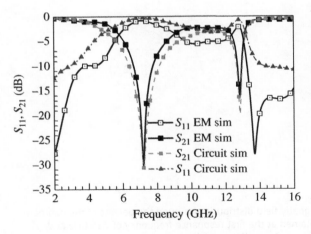

Figure 2.30 S-parameters for the equivalent circuit in Figure 2.29 compared with EM simulated data of the DGS integrated line shown in Figure 2.26. Source: Adapted from Chang et al. [13].

N_s number of DGS slots and each slots is modeled as a transmission line having characteristic impedance Z_0^{slot} and electrical length $\theta = \theta_1 + \theta_2 = \beta_m^{slot} d$. The reference of the input port is determined by the location of the microstrip line which for a symmetrical deployment with respect to the slot width "d," results in $\theta_1 = \theta_2$. The coupling between the slots and microstrip line is represented by an

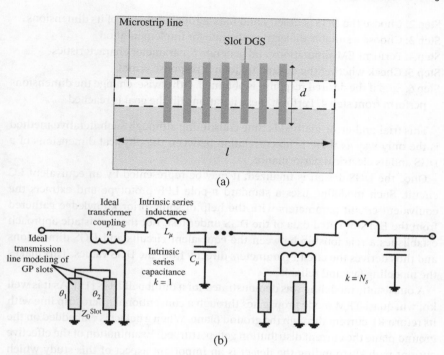

Figure 2.31 Modelling of a simple 7-cell uniform periodic slot DGS: (a) Top view; (b) Transmission line-based equivalent circuit model. Source: Adapted from Caloz et al. [14].

ideal transformer with turn ratio n expressed as

$$n \cong \sqrt{\frac{Z_0^{\mu strip}}{Z_0^{slot}}} \qquad (2.46)$$

The value of n appears to be a function of frequency. However, it varies roughly by 20% for frequencies varying from dc to 25 GHz [14]. Therefore, the value of n is chosen for some intermediate frequency.

2.5 Quasistatic Modeling

This is a unique modeling technique demonstrated in [15] for a dumbbell DGS. Its uniqueness relative to those described in Sections 2.2–2.4 could be understood from the following discussions. The conventional design approach is followed:

Step 1: Fix the desired specifications (e.g. frequency of operation, bandwidth, stopband attenuation, Q-factor, etc.)

Step 2: Choose the DGS geometry and have a rough estimate of its dimensions
Step 3: Choose a suitable dielectric substrate for implementation
Step 4: Perform EM simulations and generate *S*-parameter characteristics
Step 5: Check whether the desired frequency response is met.
Step 6: Stop if the desired response is obtained. Otherwise, change the dimensions
perform from step 4. Perform these iterations till the goal is reached.

This trial-and-error method is time consuming although such iterative method is the only way as there is no correlation between the physical dimensions of a DGS and its electrical performance.

Once the DGS design is finalized, it may be represented by an equivalent LC circuit. Such modeling uses a standard n-pole LPF prototype and extracts the equivalent circuit parameters with the help of some prior knowledge gathered from the EM simulated data of the DGS under test. But the quasistatic approach establishes a relationship between the equivalent circuits and DGS dimensions and thus derives the circuit parameters directly from the DGS values. This makes the modeling fast and reliable.

A quasistatic modeling was demonstrated for a dumbbell DGS [15]. As it is well known, quasi-TEM mode propagates through a conventional microstrip line with its return RF current through the ground plane. When a defect is embedded on the ground plane the current distribution gets perturbed. Examination of the effective current path surrounding the defect is an important aspect of this study which could be obtained by a commercial EM simulator as shown in Figure 2.32 [15].

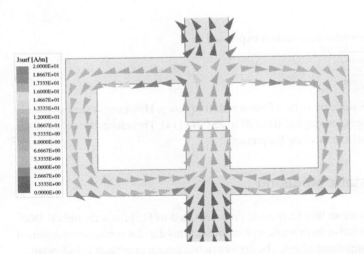

Figure 2.32 Surface current distribution around the periphery of the defect obtained using a commercial EM simulator. Source: Adapted from Karmakar et al. [15].

A compact 3D structure of gap coupled microstrip line is shown in Figure 2.33a. The current concentration areas help in portraying a 2D equivalent current ribbon model as depicted in Figure 2.33b. It reveals a pair of microstrip cross structures along with a gap discontinuity. Their individual modeling is the next step as presented below.

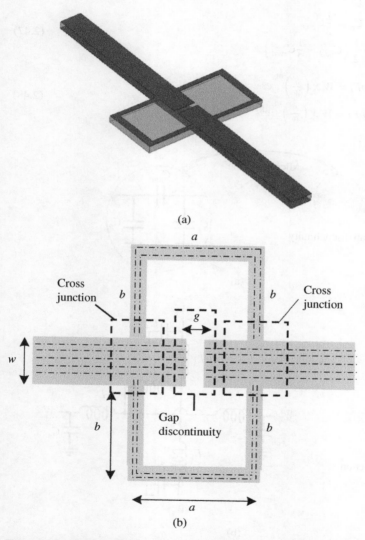

(a)

(b)

Figure 2.33 Modelling of a gap-coupled microstrip line on a dumbbell DGS: (a) 3D view; (b) equivalent current ribbon. Source: Adapted from Karmakar et al. [15].

2.5.1 Microstrip Gap Model

An equivalent circuit for a microstrip gap discontinuity is shown in Figure 2.34a [15] which consists of two parallel line capacitances C_p in conjunction with a gap capacitance C_{gap}. Their values are extracted from the odd and even mode capacitances [15], [16] as

$$\begin{cases} C_p = \dfrac{1}{2}C_{even} \\ C_{gap} = \dfrac{1}{2}\left(C_{odd} - \dfrac{1}{2}C_{even}\right) \end{cases} \tag{2.47}$$

$$\begin{cases} C_{odd}\,(pF) = W\,x\left(\dfrac{S}{W}\right)^{m_o} e^{K_o} \\ C_{even}\,(pF) = W\,x\left(\dfrac{S}{W}\right)^{m_e} e^{K_e} \end{cases} \tag{2.48}$$

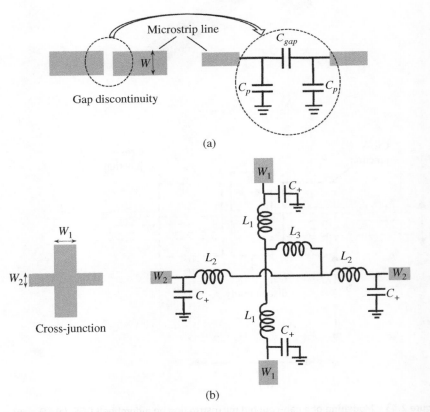

(a)

(b)

Figure 2.34 Equivalent circuits for the segments depicted in Figure 2.33: (a) Microstrip gap; (b) Microstrip cross. Source: Adapted from Karmakar et al. [15].

$$\begin{cases} m_o = \dfrac{W}{h}\left(0.619log\dfrac{W}{h} - 0.3853\right) \\[2mm] K_o = 4.26 - 1.453log\dfrac{W}{h} \end{cases} \quad \text{for } 0.1 \le \dfrac{S}{W} \le 0.3 \qquad (2.49)$$

$$\begin{cases} m_e = 0.8675 \\[2mm] K_e = 2.043\left(\dfrac{W}{h}\right)^{0.12} \end{cases} \quad \text{for } 0.1 \le \dfrac{S}{W} \le 0.3 \qquad (2.50)$$

$$\begin{cases} m_e = \dfrac{1.565}{\left(\dfrac{W}{h}\right)^{0.16}} - 1 \\[2mm] K_e = 1.97 - \dfrac{0.03}{\dfrac{W}{h}} \end{cases} \quad \text{for } 0.3 \le \dfrac{S}{W} \le 1 \qquad (2.51)$$

$$\begin{cases} C_{odd}(pF) = C_{odd}(9.6)\left(\dfrac{\varepsilon_r}{9.6}\right)^{0.8} \\[2mm] C_{even}(pF) = C_{even}(9.6)\left(\dfrac{\varepsilon_r}{9.6}\right)^{0.9} \end{cases} \qquad (2.52)$$

where S = gap width, W = microstrip line width, $\varepsilon_r = 9.6$.

2.5.2 Microstrip Cross Junction Model

The cross junction marked in Figure 2.33b has been modeled in Figure 2.34b as LC combinations. The inductance and capacitance values are calculated using a set of relations developed in [16–19] as

$$C_+ (pF) = W_1 \left\{ \left(\dfrac{W_1}{h}\right)^{-\frac{1}{3}} \left[\left(86.6\dfrac{W_2}{h} - 30.9\sqrt{\dfrac{W_2}{h}} + 367\right) log\left(\dfrac{W_1}{h}\right) \right.\right.$$
$$\left.\left. + \left(\dfrac{W_2}{h}\right)^3 + 74\dfrac{W_2}{h} + 130 \right] - 1.5\dfrac{W_1}{h}\left(1 - \dfrac{W_2}{h}\right) + \dfrac{2h}{W_2} - 240 \right\} \qquad (2.53)$$

$$L_1 = L_2 = h\left(\dfrac{W_1}{h}\right)^{-\frac{3}{2}}$$
$$\times \left[\dfrac{W_1}{h}\left(165.6\dfrac{W_2}{h} + 31.2\sqrt{\dfrac{W_2}{h}} - 11.8\left(\dfrac{W_2}{h}\right)^2\right) - 32\dfrac{W_2}{h} + 3 \right] \qquad (2.54)$$

$$L_3 = h\left[337.5 + \dfrac{h}{W_2}\left(1 + \dfrac{7h}{W_1}\right) - 5\dfrac{W_2}{h}cos\left(\dfrac{\pi}{2}\left(1.5 - \dfrac{W_1}{h}\right)\right) \right] \qquad (2.55)$$

where W_1 and W_2 are the microstrip line widths, h = thickness

2.5.3 Modeling of the Rest Current Paths

The current paths marked as *a* and *b* in Figure 2.33b form a U-shaped current filament which is very logically modeled as an indicator. The inductances are calculated in [16–22]. The complete equivalent circuit of a dumbbell DGS is now shown in Figure 2.35. The equivalent circuit parameters are directly related to the physical parameters of both substrate and DGS. The accuracy of the equivalent circuit has been examined in Figure 2.36 as the functions of DGS dimensions. The EM simulation has been considered as the reference for comparison indicating an excellent mutual agreement. Some degree of deviation has been observed in predicting the operating bandwidth. The reason is the mode of calculation which is valid for narrowband approximations.

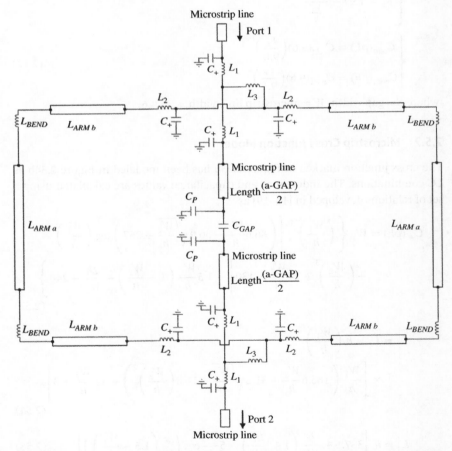

Figure 2.35 Equivalent circuit of a unit cell dumbbell-shaped DGS based on quasistatic model. Source: Karmakar et al. [15] © [2006] IEEE.

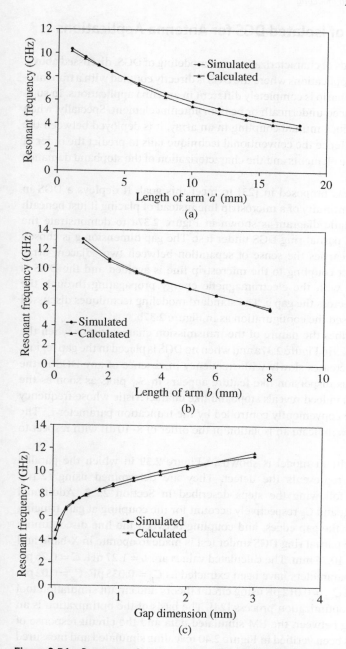

Figure 2.36 Resonance frequency of the dumbbell DGS predicted by quasi static modelling-based calculation and EM simulated data: (a) variation as a function of equivalent ribbon length "a" (Figure 2.33); (b) variation as a function of equivalent ribbon length "b" (Figure 2.33); (c) variation as a function of the gap dimension "g" (Figure 2.33). Source: Karmakar et al. [15] © [2006] IEEE.

2.6 Modeling of Isolated DGS for Antenna Applications

The standard methods of characterization and modeling of DGS, discussed above, are ideal for circuit applications where the DGS is directly coupled with a transmission line. But the scenario is completely different in antenna applications. In such cases, the DGS is placed underneath or near an antenna element. Specially, when a DGS is used to reduce mutual coupling in an array, it is deployed between the adjacent elements. Hence the conventional technique fails to predict the order of isolation between the elements and the characterization of the stopband demands a different approach.

A new concept was proposed in [23] to meet this goal. It deploys a DGS in between a gap discontinuity of a microstrip line instead of placing it just beneath the same. A schematic diagram is shown in Figure 2.37a to demonstrate the technique [23] with partial ring DGS under test. The gap dimension g is chosen such a way that it carries the sense of separation between two adjacent array elements. Any direct coupling to the microstrip line is avoided and the DGS is allowed to interact with the electromagnetic energy propagating through the dielectric medium across the gap g. The standard modeling techniques discussed so far would have used the configuration as in Figure 2.37b.

Figure 2.38 describes the nature of the transmission characteristics when the DGS is gap-coupled as in Figure 2.37a and when no DGS is placed in the gap region. The value of S_{21} increases slowly with frequency in absence of any DGS in the line-gap. Anomalous dispersion like feature appears in S_{21} plots as soon as the DGS is placed. This indeed reveals stopband like characteristic whose frequency of operation can be conveniently controlled by the truncation parameter t_r. The simulated S_{21} values indicate an isolation of the order of 8–10 dB with respect to the "no-DGS" case.

The equivalent circuit model is shown in Figure 2.39 in which the parallel LCR combination represents the defect. They are determined using (2.10), (2.11), and (2.14) following the steps described in Section 2.2.3. Additional capacitances C_g, C_p, and C_c respectively account for the coupling at gap, fringing field coupling near the gap edges, and coupling between the line discontinuity and the defect. The partial ring DGS under test is made to operate in X-band and corresponding $t_r = 10.75$ mm. The calculated values are $L = 1.27$ nH, $C = 0.26$ pF, $R = 600\,\Omega$. Other parameters have been extracted as $C_p = 0.035$ pF, $C_g = 0.02$ pF, $C_{c1} = 0.012$ pF, and $C_{c2} = 0.012$ pF using circuit theory and circuit simulation tool through a built-in optimization process [24]. The basis of the optimization is an optimum matching between the EM simulated data and the circuit response of S_{21}. This model has been verified in Figure 2.40 revealing simulated and measured data with close mutual agreement. This modeling technique has been used to design partial ring DGS to demonstrate suppression of mutual coupling between

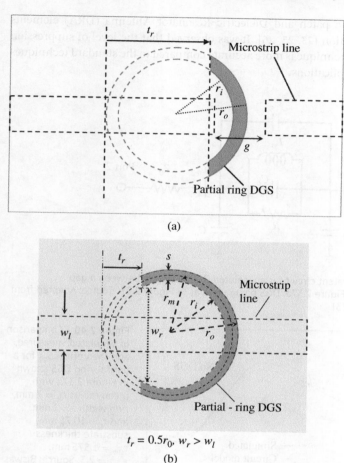

(a)

$t_r = 0.5r_0,\ w_r > w_l$

(b)

Figure 2.37 (a) Partial ring DGS placed at the middle of a gap discontinuity of a microstrip line. (b) A partial-ring DGS deployed underneath a microstrip line. Source: Adapted from Biswas and Guha [23].

Figure 2.38 Simulated S_{21} for the configuration in Figure 2.37a with and without DGS. Parameters: $g = 5\,\text{mm}$, $s = 1\,\text{mm}$, $r_m = 7\,\text{mm}$, $t_r = 3.75\,\text{mm}$, 7.5 mm, and 10.75 mm, substrate thickness $t_{sub} = 1.575\,\text{mm}$, $\varepsilon_{rsub} = 2.3$. Source: Biswas and Guha [23] © [2013] PIERS.

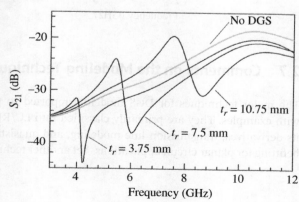

adjacent microstrip patch and Dielectric Resonator Antenna (DRA) elements in array configuration [23, 25, 26]. It was observed that the level of suppression predicted by this technique is more accurate compared to the standard techniques used for circuit applications.

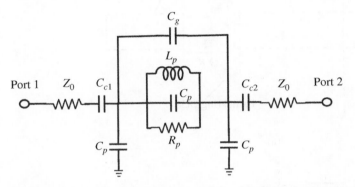

Figure 2.39 Equivalent circuit for an isolated DGS placed in between a gap discontinuity as in Figure 2.37a. DGS parameters as in Figure 2.38. Source: Adapted from Biswas and Guha [23].

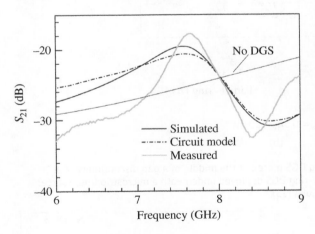

Figure 2.40 Comparison of simulated, measured, and theoretical S_{21} for a partial-ring DGS shown in Figure 2.37a with mean radius $r_m = 7$ mm, ring width $s = 1$ mm, and $t_r = 10.75$ mm, substrate thickness $t_{sub} = 1.575$ mm, $\varepsilon_{rsub} = 2.3$. Source: Biswas and Guha [23] © [2013] PIERS.

2.7 Comments on the Modeling Techniques

The major techniques for DGS modeling reported so far have been discussed with examples. They are primarily classified into LC/RLC circuit modeling and its derivatives, transmission line modeling, and quasistatic approach which are befitting for planar circuit applications. Yet another technique has been described which deals with an isolated DGS and is ideal for printed antenna and array applications.

Among the three major modeling for circuit applications, the quasistatic approach appears to be more practical. In this method, the physical dimensions of a DGS are directly related to the equivalent circuit elements. This allows one to design a DGS purely based on a theoretical approach. But LC/RLC technique lacks in this aspect since one needs to guess the DGS dimensions according to the given frequency response and then to follow iterative EM simulations until the desired response is achieved. This is time consuming and not that inefficient.

The quasistatic model, as mentioned above, is advantageous compared to the LC/RLC modeling, but it is fully a geometry dependent method. The given example of a dumbbell DGS could easily be handled since its geometry is realizable in terms of a microstrip gap and microstrip cross structures whose equivalent circuits are readily available from some earlier works. Therefore, it may not be readily applicable to any arbitrary DGS shape as per our choice.

Hence, the LC circuit modeling with some advanced modifications seems to be the best option. The transmission line modeling also provides a better insight but its dispersive nature limits the flexibility and accuracy. For antenna applications, the modeling depends on the deployment of the defect and principle of operation. A designer, therefore, would be the best one to judge the requirement and hence choose the best possible technique to model the DGS of interest.

References

1 D. Ahn, J. S. Park, C. S. Kim, J. Kim, Y. Qian, and T. Itoh, "A design of the low-pass filter using the novel microstrip defected ground structure," *IEEE Transactions on Microwave Theory and Techniques,* vol. 49, no. 1, pp. 86–93, 2001.

2 J. S. Hong and M. J. Lancaster, "Microstrip Filters for RF/Microwave Applications," 2nd Edition, *John Wiley & Sons Ltd.*, 2011.

3 I. Chang and B. Lee, "Design of defected ground structures for harmonic control of active microstrip antenna," *IEEE Antennas and Propagation Society International Symposium,* vol. 2, pp. 852–855, 2002, San Antonio, TX, USA.

4 J. S. Park, J. H. Kim, J. H. Lee, S. H. Kim, and S. H. Myung, "A novel equivalent circuit and modeling method for defected ground structure and its application to optimization of a DGS lowpass filter," *IEEE MTT-S International Microwave Symposium Digest,* vol. 1, pp. 417–420, 2002, Seattle, WA, USA.

5 C. S. Kim, J. S. Lim, S. Nam , K. Y. Kang, and D. Ahn, "Equivalent circuit modelling of spiral defected ground structure for microstrip line," *Electronics Letters*, vol. 38, no. 19, pp. 1109–1110, 2002.

6 C.-S. Kim, J.-S. Lim, S. Nam, K.-Y. Kang, J.-I. Park, G.-Y. Kim, and D. Ahn, "The equivalent circuit modeling of defected ground structure with spiral shape," *2002 IEEE MTT-S International Microwave Symposium Digest*, vol. 3, pp. 2125–2128, 2002, Seattle, WA, USA.

7 M. Makimoto and S. Yamashita, "Bandpass filters using parallel coupled stripline stepped impedance resonators," *IEEE Transactions on Microwave Theory and Techniques*, vol. 28, no. 12, pp. 1413–1417, 1980,

8 J.-S. Hong and B. M. Karyamapudi, "A general circuit model for defected ground structures in planar transmission lines," *IEEE Microwave and Wireless Components Letters*, vol. 15, no. 10, pp. 706–708, 2005.

9 D. A. Frickey, "Conversions between S, Z, Y, H, ABCD, and T parameters which are valid for complex source and load impedances," *IEEE Transactions on Microwave Theory and Techniques*, vol. 42, no. 2, pp. 205–211, 1994.

10 A. Balalem, A. R. Ali, J. Machac, and A. Omar, "Quasi-elliptic microstrip low-pass filters using an interdigital DGS slot," *IEEE Microwave and Wireless Components Letters*, vol. 17, no. 8, pp. 586–588, 2007.

11 G. Matthei, L. Young, and E. M. T. Jones, *"Microwave Filters, Impedance Matching Networks, and Coupling Structures,"* Norwood, Artech House, 1980.

12 J.-X. Chen, J.-L. Li, K.-C. Wan, and Q. Xue, "Compact quasi-elliptic function filter based on defected ground structure," *IEE Proceedings – Microwave, Antennas and Propagation*, vol. 153, no. 4, pp. 320–324, 2006.

13 C.-C. Chang, C. Caloz, and T. Itoh, "Analysis of a compact slot resonator in the ground plane for microstrip structures," *2001 Asia-Pacific Microwave Conference*, vol. 3, pp. 1100–1103, 2001, Taipei, Taiwan.

14 C. Caloz, H. Okabe, T. Iwai, and T. Itoh, "A simple and accurate model for microstrip structures with slotted ground plane," *IEEE Microwave and Wireless Components Letters*, vol. 14, no. 4, pp. 133–135, 2004.

15 N. C. Karmakar, S. M. Roy, and I. Balbin, "Quasi-static modeling of defected ground structure," *IEEE Transactions on Microwave Theory and Techniques*, vol. 54, no. 5, pp. 2160–2168, 2006.

16 R. Garg and I. J. Bahl, "Microstrip discontinuities," *International Journal of Electronics*, vol. 45, no. 1, pp. 81–87, 1978.

17 E. J. Denlinger, "A frequency dependent solution for microstrip transmission lines," *IEEE Transactions on Microwave Theory and Techniques*, vol. MTT-19, no.1, pp. 30–39, 1971.

18 A. Gopinath and P. Silvester, "Calculation of inductance of finite—Length strips and its variation with frequency," *IEEE Transactions on Microwave Theory and Techniques*, vol. MTT-21, no. 6, pp. 380–386, 1973.

19 A. F. Thomson and A. Gopinath, "Calculation of microstrip discontinuity inductances," *IEEE Transactions on Microwave Theory and Techniques*, vol. MTT-23, no. 8, pp. 648–655, 1975.

20 F. W. Grover, "*Inductance Calculation: Working Formulas and Tables*," Dover, New York, 1946.

21 B. Easter, "The equivalent circuit of some microstrip discontinuities," *IEEE Transactions on Microwave Theory and Techniques*, vol. MTT-23, no. 8, pp. 655–660, 1975.

22 B. L. Ooi, D. X. Xu, and L. H. Guo, "Efficient methods for inductance calculation with special emphasis on nonuniform current distributions," *Microwave and Optical Technology Letters*, vol. 40, no. 4, pp. 432–436, 2004.

23 S. Biswas and D. Guha, "Isolated open-ring defected ground structure to reduce mutual coupling between circular microstrips: characterization and experimental verification," *Progress In Electromagnetics Research M*, vol. 29, pp. 109–119, 2013.

24 *PathWave Advanced Design System (ADS)*.

25 D. Guha and S. Biswas, "Characterization of defected ground structure to be used between two DRA array elements for suppressing the mutual coupling," *Proceedings of the 2012 IEEE International Symposium on Antennas and Propagation*, pp. 1–2, 2012, Chicago, IL, USA.

26 S. Biswas and D. Guha, "Stop-band characterization of an isolated DGS for reducing mutual coupling between adjacent antenna elements and experimental verification for dielectric resonator antenna array," *AEU – International Journal of Electronics and Communications*, vol. 67, no. 4, pp. 319–322, 2013.

3

DGS for Printed Antenna Feeds

3.1 Introduction

Historically, the need of periodic slots underneath a microstrip feed line to an antenna dates back to 1998 [1]. That time, the arrangement of periodic defects was commonly called as photonic band-gap (PBG) structure, which was renamed subsequently as electromagnetic band-gap (EBG). However, in [1], the purpose of EBG was to filter out the higher harmonics produced by a power amplifier feeding to a slot radiator. A contemporary work [2] reported an identical mechanism where the radiating element was a microstrip patch. So called PBG or EBG structures used in [1, 2] actually exploited the reactive impedance of the defects resulting in a filtering property with a passband. The radiating elements were not under any consideration in those designs.

The evolution of DGS has been discussed in Chapter 1. In place of a widely spread periodic PBG or EBG patterns, a few limited units of DGS came into operation around 2000 [3, 4] and they were explored to meet several challenges in planar feed designs. They include matching of the input impedance, elimination of harmonics, increase in isolation between the ports, and control of the excitation phase. They are truly significant and are becoming more and more relevant in the modern complex and compact mobile devices. Cost effectiveness along with low profile feature makes DGS commercially viable. A comprehensive account of such developments and applications has been covered in this chapter.

3.2 Impedance Matching of Antenna Feed Lines

Matching of the input impedance of an antenna with its feed is one of the most vital considerations of antenna engineering. Because the input impedance widely varies with the type and shape of an antenna along with the feed location. A good example is a microstrip patch whose impedance typically varies from $0\,\Omega$ at the

Defected Ground Structure (DGS) Based Antennas: Design Physics, Engineering, and Applications,
First Edition. Debatosh Guha, Chandrakanta Kumar, and Sujoy Biswas.

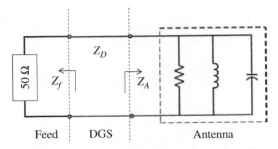

Figure 3.1 Schematic circuit diagram of an antenna feed: DGS beneath the feed line is used as impedance matching network [5]. Source: Adapted from Sung and Kim [5].

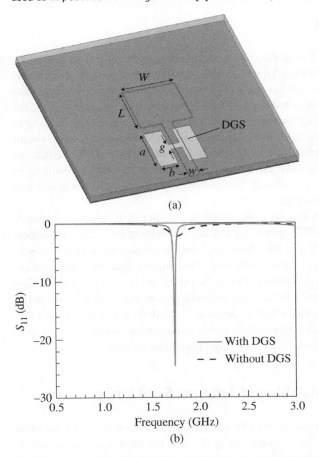

Figure 3.2 An edge fed rectangular patch fed by a 50 Ω line where "H"-shaped DGS provides the required impedance matching: (a) schematic view (b) simulated S_{11} versus frequency. $L = W = 56$ mm, $a = 18.8$ mm, $b = 7$ mm, $g = 0.4$ mm, $w = 4.7$ mm. Source: Adapted from Sing and Kim [5].

center to about 300 Ω at the radiating edge. But the characteristic impedance of a standard feed line lies between 30 and 100 Ω. A designer, therefore, needs either inset feeding or an additional section of matching network which in turn requires additional deployment area. The space constraint in a compact system or in an array may not allow the same. But this stringent requirement can be technically accomplished by integrating DGS(s) with a printed feed line. A scheme is described in Figure 3.1 [5] which is self-explanatory. Feed impedance Z_f is typically 50 Ω for a commercial feed or connector and the antenna impedance Z_A comprises a finite resistance along with a non-zero reactive component. The intermediate section represents a microstrip line connecting the antenna with a commercial feed. It can be considered as a two-port network whose image impedance [6] can be easily controlled by a DGS. The shape and size of the DGS help in achieving an effective line impedance Z_D which satisfies the required impedance matching.

One such application is shown in Figure 3.2a where a square patch etched on 1.575 mm thick substrate with $\varepsilon_r = 2.2$ is fed by a 50 Ω microstrip line and the specified frequency of operation is 1.75 GHz. The feed point resistance near patch edge is about 300 Ω and such high impedance is achieved by introducing a H-shaped DGS underneath the line [5]. The working principle of such DGS beneath a printed line has already been discussed in the previous chapters. The S_{11} characteristics of the antenna with and without DGS are depicted in Figure 3.2b. They signify the role of the DGS that helps in achieving a perfect match with $S_{11} \approx -25$ dB at the design frequency of 1.75 GHz.

Such requirements may sometimes be very stringent. One such example is shown in Figure 3.3 [7]. It shows a microstrip fed circularly polarized patch where the DGSs lead a role in achieving impedance matching at all inputs. The patch is orthogonally fed by a 90° branch-line coupler with 50 Ω input ports. The output ports are connected to the patch edges with high impedance. To meet the matching requirement, it follows the same design principle as in Figure 3.1 and uses two H-shaped DGSs below two respective feed sections. Those DGSs are visible from the back side as shown in Figure 3.3b. The final matching is reflected by the VSWR measured at one input port as depicted in Figure 3.3c.

3.3 Controlling the Harmonics in Printed Antennas

The modern communication systems prefer a common platform for integrating antennas with the circuits such as oscillator and amplifiers. In the transmit mode at frequency f_0, there is every possibility of facing the higher harmonics such as $2f_0$ and $3f_0$, caused by the nonlinearity factor of the integrated oscillators and amplifiers. But no such out-of-band radiations is acceptable. On the other hand, in

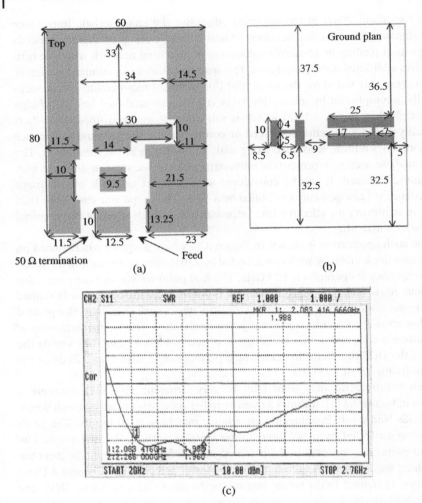

Figure 3.3 Circularly polarized near square patch integrated with DGS for impedance matching (a) top view (b) view from ground plane side. NOTE: area in white mean metalized section and in gray means the etched-out area or substrate, (c) VSWR response with DGS [7]. Source: Thakur and Park [7] © [2006] IEEE.

the receive mode, any unwanted signals beyond the specified band of frequencies may saturate the low noise amplifier (LNA). Hence a suitable mechanism needs to be adopted to get rid of such occurrences. Indeed, the resonant property of a DGS has been explored since 1998 [1] as an effective and viable solution to this issue. Several approaches have been tried subsequently which encompass three categories depending on the order of harmonics addressed. They have been a little elaborated in the following discussions.

3.3.1 Suppression of Second Harmonic ($2f_0$)

A 2D pattern of circular defects was used below a microstrip line for the first time in [1] to feed an antenna. The arrangement is shown in Figure 3.4. A rectangular

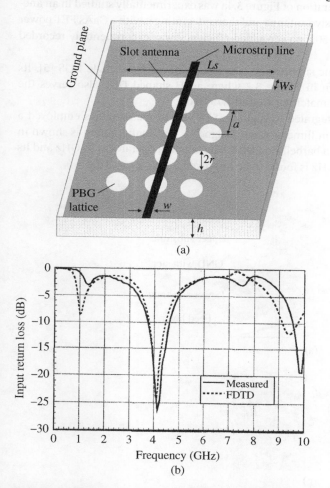

(a)

(b)

Figure 3.4 50 Ω Feed line to a slot antenna on 2D PBG lattice [1]: (a) isometric view, (b) S_{11} response indicating suppressed 2nd harmonic around 8 GHz. Substrate: 0.79 mm thick RT/Duroid 5870 with $\varepsilon_r = 2.33$, defect radius 2.8 mm, lattice period $a = 12.2$ mm, radiating slot dimension: 55.88 mm × 3.3 mm [1]. Source: Radisic et al. [1] © [1998] IEEE.

slot is fed by a 50 Ω microstrip line and a 4 × 3 array of circular defects has been created on the ground plane below the feed. As discussed earlier, such 2D array of defects was commonly called "PBG lattice" which subsequently evolved as "DGS" [3] in the form of limited number of defects. The antenna described in Figure 3.4a hardly shows any trace of its 2nd harmonic as evident from the measured and computed S_{11} traces shown in Figure 3.4b. The second harmonic of the fundamental resonance at 4.2 GHz is clearly absent near 8.4 GHz.

The antenna configuration of Figure 3.4a was experimentally studied in an anechoic chamber as a passive antenna and then integrating it with a GaAs FET power amplifier. Better than 50% power-added efficiency was experimentally recorded over 3.7 to 4.0 GHz.

The second harmonic suppression was also achieved by a single DGS [5]. Its configuration is shown in Figure 3.2 where an H-shaped DGS also serves the purpose of impedance matching around 1.75 GHz.

A 2 × 2 substrate integrated waveguide (SIW)-based rectenna has employed a similar approach in a millimeter wave design [8]. The configuration is shown in Figure 3.5 which uses a barbell ring DGS. The antenna operates at 35 GHz and its S_{11} response near 70 GHz is found to be improved from −8 to −3 dB.

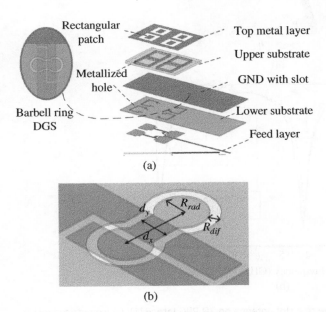

(a)

(b)

Figure 3.5 Barbell ring DGS integrated rectifying SIW antenna at 35 GHz: (a) antenna configuration, (b) enlarged view of the DGS. DGS parameters: $R_{rad} = 0.25$ mm, $R_{dif} = 0.1$ mm, $d_x = 0.72$ mm, $d_y = 0.3$ mm [8]. Source: Wang et al. [8] © [2021] IEEE.

3.3.2 Suppression up to Third Harmonic ($3f_0$)

After [1], a contemporary work also applied 2D lattice of circular defects [2], the layout being shown in Figure 3.6a [2]. The patch etched on a 1.6 mm thick glass epoxy substrate ($\varepsilon_r = 4.8$) operates near $f_0 = 900$ MHz. The array of defects, each measuring 0.054 λ_0 diameter, is a bit differently deployed spreading under both feed line and radiating patch. Interestingly, the reason behind placing the defects under the patch is not clear, rather they drastically affect the antenna gain. But the feature of harmonic suppression up to $3f_0$ has been well established through the studies in Figure 3.6b. It allows 900 MHz signal to propagate smoothly but neither

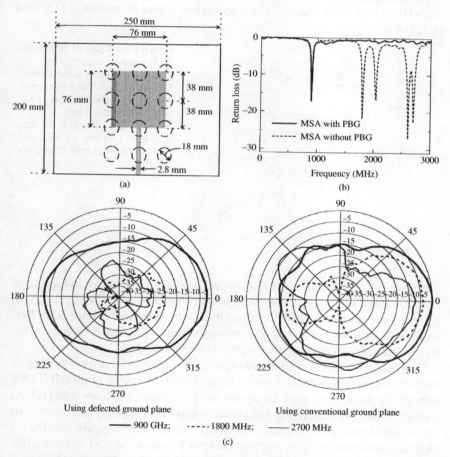

Figure 3.6 Suppression of harmonic radiations from a microstrip patch using an array of circular DGSs: (a) antenna configuration, (b) measured and simulated reflection coefficient, (c) measured H-plane radiation patterns. [2]. Source: Horii and Tsutsumi [2] © [1999] IEEE.

the second harmonic ($2f_0 = 1800$ MHz) nor the third harmonic ($3f_0 = 2700$ MHz). This is ultimately manifested through the radiation characteristics as compared in Figure 3.6c. The radiation patterns at 900 MHz remains unaffected except a gain reduction by about 2 dB. The defects near the patch boundary are not at all healthy in terms of fringing field distribution and hence the effective radiating aperture causing decrement in gain. However, the radiations corresponding to the second and third harmonics get significantly suppressed by about 15 and 18 dB, respectively.

An advanced and useful design was reported in [9]. It is relatively simpler as shown in Figure 3.7 [9] that radiates at 5.775 GHz. A pair of dumbbell-shaped DGSs [10] has been used in series to suppress two higher harmonics at 11.55 and 17.325 GHz.

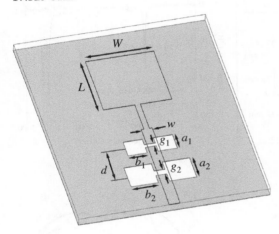

Figure 3.7 A pair of dumbbell DGSs used for suppressing harmonic radiations. Substrate FR4 with $\varepsilon_r = 4.7$ and thickness 1.6 mm, $L = W = 11.3$ mm, $w = 2.2$ mm, $d = 3$ mm, $a_1 = b_1 = 1.45$ mm $g_1 = 0.4$ mm, $a_2 = b_2 = 1.94$ mm, $g_2 = 0.3$ mm. Source: Adapted from Chang and Lee [9].

An interesting combination of PBG lattice (used in [2] and shown in Figure 3.6) and DGS (used in [9] and shown in Figure 3.7) has been examined in [11]. That geometry is depicted in Figure 3.8a and its S_{11} characteristics (Figure 3.8b) look very similar to those in Figure 3.6b. Only one noticeable change is revealed in the form of reduction in ripple in the stopband by about 0.5 dB.

The H-shaped DGS [5] as discussed in previous chapters has been explored again for harmonic suppressions in a very compact form [12]. It uses a unit cell H-DGS beneath the neck of an inset-fed square patch as shown in Figure 3.9a [12]. As a result, it does not require any extra space beyond the periphery of the patch. Its fundamental resonance occurs at 1.82 GHz as is revealed from Figure 3.9b, but the DGS integrated design takes care of the higher harmonics at 3.64 and 5.46 GHz. This records up to 20 dB suppression in the radiated fields at the second harmonic and that too weak to measure at the third harmonic [12]. This design features significant reduction in the etched-out region under the patch compared to that in

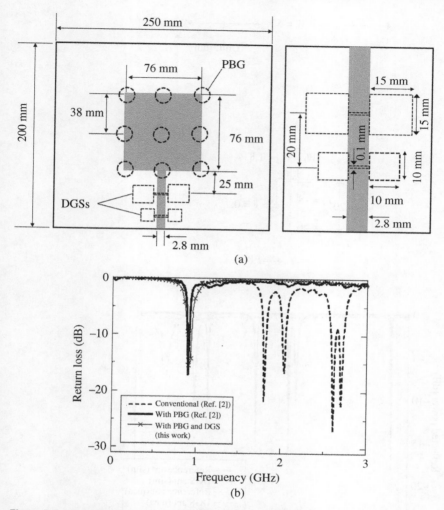

Figure 3.8 Combination of circular and dumbbell shaped DGSs for wider stopband: (a) antenna geometry, with magnified view of DGS shown at right side (b) S-parameter under different conditions [11]. Source: Liu et al. [11] © [2005] IEEE.

Figure 3.6 [2]. This in turn results in about 5 dB relative improvement in the front to back ratio of the radiated fields.

A wide stopband has been reported by using a single unit of inverted U-shaped DGS [13]. It is integrated with a 2.45 GHz ring slot antenna as shown in Figure 3.10. The stopband starts from 3 GHz and continues up to about 10 GHz. This configuration works well for other shapes of ring slots [13].

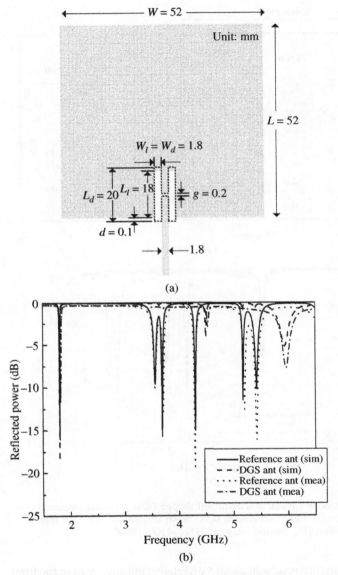

(a)

(b)

Figure 3.9 H-shaped DGS near the neck of the feed to patch junction: (a) antenna configuration, (b) comparison of S_{11} with reference antenna. Substrate: $\varepsilon_r = 2.2$, thickness 0.508 mm [12]. Source: Sung et al. [12] © [2003] IEEE.

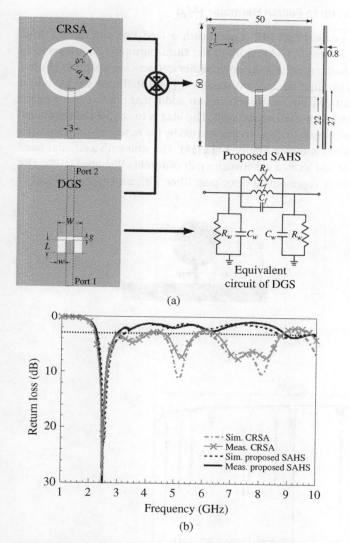

Figure 3.10 Circular slot antenna integrated with inverted U-shaped DGS for harmonic suppression: (a) antenna configuration, (b) S_{11} characteristics. $a_1 = 10$, $a_2 = 13.5$, $L = 8$, $w = 2.25$, and $g = 0.5$, all dimensions in mm [13]. Source: Sim et al. [13] © [2014] IEEE.

3.3.3 Suppression up to Fourth Harmonic ($4f_0$)

Typically, a single or dual cell DGS underneath a feed line can offer a wide stopband but with a limit of tackling up to the third harmonic as discussed in Section 3.3.2. The challenge of chasing up to higher frequencies, say up to $4f_0$, has also been reported through several investigations [14–16]. Those works indeed innovated a different concept by introducing an additional reactive or resonant structure in the form of an open-ended stub. The idea is to couple the resonance characteristics of one or two DGSs with that caused by the newly added open stub. One such configuration is shown Figure 3.11a [14]. The stub with a circular head is attached to the line and located in between a pair of dumbbell-shaped DGSs. The filter section effectively represents a three pole filter with a steeper passband to

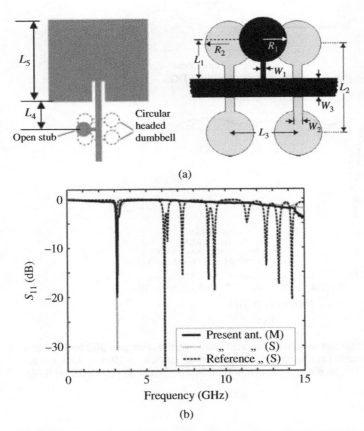

Figure 3.11 Feed with DGS integrated filter for suppressing up to the 4th harmonic: (a) antenna layout, (b) S_{11} versus frequency. Substrate 0.381 mm thin RT duroid 5880 ($\varepsilon_r = 2.2$), parameters: $W_1 = W_2 = 0.5$, $W_3 = 1.2$, $L_1 = 3.4$, $L_2 = 6$, $R_1 = R_2 = 2$, $L_4 = 18.5$, $L_5 = 32$, all dimensions in mm [14]. Source: Mandal et al. [14] John Wiley & Sons.

stopband transition compared to any three pole DGS filter [14]. The S_{11} response of the antenna furnished through Figure 3.11b ensures an excellent rejection up to about 15 GHz leaving the fundamental resonance at 3.12 GHz unaffected.

The filter section actually occupies some space around the feed and that can be made more compact as demonstrated in [15]. The above principle has been followed in [15] by reconfiguring the DGS geometry. It uses a pair of partial ring-shaped DGSs as shown in Figure 3.12 [15] revealing very similar performance

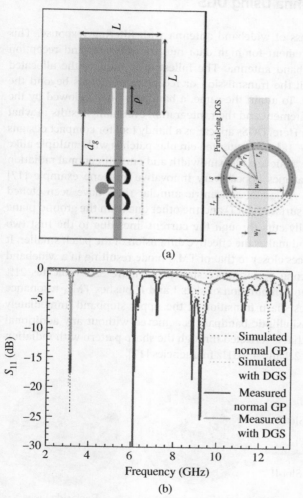

Figure 3.12 Microstrip patch integrated with a filter section comprising of partial ring DGSs and an open stub: (a) antenna geometry, (b) reflection coefficient versus frequency of the antenna. Substrate: 0.508 mm thin RT Duroid 5880 ($\varepsilon_r = 2.2$); parameters $w_r = w_1 = 1.6$, $s = 0.5$, $r_m = 2.5$, $L = 32$, $\rho = 11.6$, $d_g = 11.62$, all dimensions in mm [15]. Source: (a) Biswas et al. [15] © [2013] IEEE, (b) Adapted from Biswas et al. [15].

as in Figure 3.11b [14]. Interestingly, this design [15] results in about 40% reduction in the size of the filter section compared to that in Figure 3.11 [14]. The measured radiation ensures a minimum of 15 dB suppression in gain for the higher order modes. A similar feed has been examined for a 3 GHz dielectric resonator antenna [16]. All higher order modes up to 12 GHz have been adequately suppressed.

3.4 Filtering Antenna Using DGS

Filtering antenna is a class of wideband antenna with filtering response. This satisfies a specific requirement for high data rate transmission and reception uniformly using a broadband antenna. The fullest utilization of the allocated band would be possible if the transmission or reception of signals beyond the band is sharply curtailed. To attain the same, a bandpass filter followed by the antenna is the basic requirement and their integration eventually results in what is called *filtering antenna*. Here, DGSs appear as a handy tool for compact designs [17, 18]. The works in [17, 18] use center fed circular patches with multiple alike modes packed together to achieve wide bandwidth and omnidirectional radiation patterns. The design strategies appear very innovative and one example [17] is shown in Figure 3.13. A pair of concentric annular rings have been etched out—one from the patch surface (slot I) and the other one from the ground plane (slot II). The DGS actually cuts through the current lines due to the first two modes TM_{01} and TM_{02} and makes the effective dimension of the patch smaller. It also brings their resonances closer to that of TM_{03} mode resulting in a wideband operation [17]. The antenna performance shown in Figure 3.14 indicates 21% matching bandwidth. The combination of slots I and II pushes TM_{04} resonance far off and thus improves sharp transition at the upper stopband immediately after TM_{03} mode. A quasi-elliptic bandpass is achieved without any additional circuit components and that is reflected through the sharp pattern with radiation minima below −30 dB at 2.9 and 4.9 GHz frequencies [17].

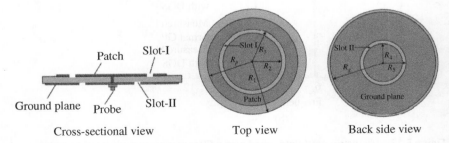

Figure 3.13 An omnidirectional filtering patch antenna integrated with annular ring DGS (slot II) [17]. Source: Wu et al. [17] © [2017] IEEE.

Figure 3.14 Simulated and measured response of the antenna shown in Figure 3.13: (a) S_{11} versus frequency, (b) realized gain. Substrate $\varepsilon_r = 2.65$, thickness 2 mm; parameters: $R_1 = 80.7$, $R_2 = 48$, $R_3 = 49.9$, $R_4 = 28.1$, $R_5 = 29$, $R_p = 68.2$, and $R_g = 79.8$ all dimension in mm [17]. Source: Wu et al. [17] © [2017] IEEE.

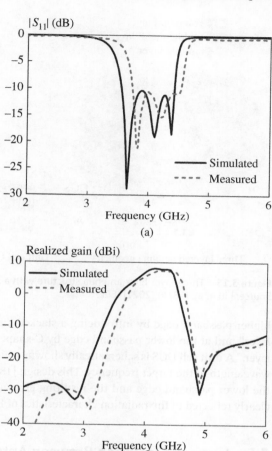

A relatively complicated but interesting design has been presented in [18]. It is basically a three layered structure as shown in Figure 3.15. The "middle layer" bears a circular patch which is the actual radiating element. A pair of concentric annular slots has been incorporated on its surface to combine quasi TM_{02}, TM_{03}, and TM_{04} modes together. A number of strategic shorting pins help TM_{01} mode to excite close by and achieve as much as 34% bandwidth. The radial slots on the patch surface are actually meant for compensating the inductive loading caused by the annular slots and thus they help in obtaining good impedance matching. This middle layer placed on a conventional ground plane results in the performance as in Figure 3.16a. A wideband response is achieved but adequate band stop characteristic is missing. That feature is achieved at the

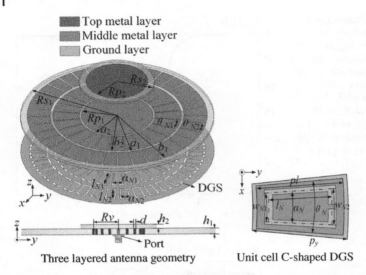

Three layered antenna geometry   Unit cell C-shaped DGS

Figure 3.15 Three layer filter antenna structure with a series of C-shaped DGSs [18]. Source: Liu et al. [18] © [2021] IEEE.

higher passband edge by introducing a stacked circular patch at the "top metal layer" and at the lower passband edge by C-shaped DGSs arrays at the "ground layer." A unit cell DGS is schematically shown in Figure 3.15 which measures half wavelength at the target frequency. This design [18], as mentioned above, targets the lower passband edge and the resulting change in filter antenna response is clearly reflected in the radiation characteristics of Figure 3.16b.

3.5 Improved Isolation Between Antenna Ports

Isolation between two ports becomes a necessary requirement when one single patch is used for both transmit and receive purposes at two different frequencies like satellite uplink and downlink. The receiver should not get saturated by its own transmitted signal and at the same time, the transmitter should not radiate any spurious signal at the receiving frequencies. A filter associated with the antenna feed can take care of this. A DGS beneath the feed line can be a good alternative to any additional filter circuit in a compact portable system [19, 20]. Figure 3.17 is an ideal example which shows a dual fed dual frequency rectangular patch with mutually orthogonal polarization. It operates at 2.5 GHz in transmit (Tx) and at 2.0 GHz in receive (Rx) modes. The LNA connected to the Rx port is protected from any 2.5 GHz high power leakage signal that may get coupled from the Tx side. A spiral-shaped DGS beneath the receiving port serves as a notch filter centered around 2.5 GHz and protects the LNA [20]. A signature of an improvement in isolation by 20 dB at 2.5 GHz is evident from the

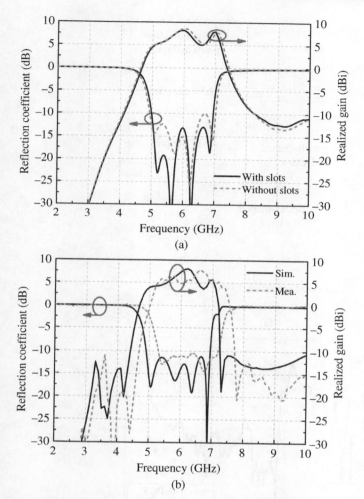

Figure 3.16 S_{11} and realized gain versus frequency of the antenna shown in Figure 3.15: (a) antenna without top metal patch and DGS, (b) proposed antenna with top metal patch and DGS as sketched in Figure 3.15. The gain was measured at 24° elevation with 0° azimuth [18]. Source: Liu et al. [18] © [2021] IEEE.

measured S_{21} shown in Figure 3.17b. It hardly affects the signal flow at 2.0 GHz. Again the power amplifier connected to the Tx port may generate a few higher harmonics resulting in reduced efficiency. Such probability has been eliminated by integrating a pair of dumbbell-shape DGS. The suppression of signals beyond 2.5 GHz eventually results in an improvement in efficiency by 3% [20].

The design approach of [20] has been followed in [21] with different geometries for both radiating patch and defects. A schematic view is shown in Figure 3.18 [21] which is meant for a full duplex omnidirectional 4G antenna. It transmits

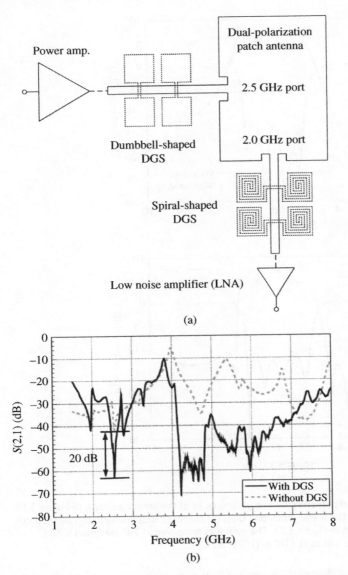

(a)

(b)

Figure 3.17 DGS used for improved isolation between the ports of a dual fed antenna: (a) antenna layout, (b) *S*-parameter to measure the isolation between two ports [20]. Source: Chung et al. [20] © [2004] IEEE.

at 3.3 GHz and receives at 3.215 GHz. Optimally designed back-to-back Ω-shaped DGSs under the feed lines result in port isolation by 42 dB at the Tx end and about 46 dB at the Rx terminal.

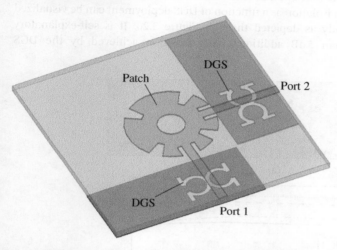

Figure 3.18 Schematic view of a DGS integrated full-duplex patch antenna on 1.6 mm FR4 Substrate to be operated in *S*-band. Source: Hussein et al. [21] John Wiley & Sons.

The DGS-based port isolation technique has been successfully extended to SIW antennas [22]. A quarter mode SIW antenna is shown in Figure 3.19 which operates in dual circularly polarized (CP) mode [22]. Port-1 and Port-2, respectively,

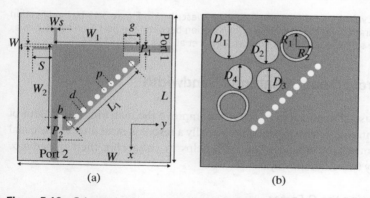

(a)   (b)

Figure 3.19 Schematic diagram of a SIW-based dual CP antenna using DGS [22]. $W = L = 23.787$, $L_1 = 14$, $S = 2.8$, $W_1 = 12.582$, $W_2 = 12.582$, $Ws = 0.5$, $g = 2.5$, $W_4 = 0.34$, $P_1 = 1.137$, $P_2 = 1.137$, $b = 0.7$, $d = 1$, $p = 1.5$, $D_1 = 5.8$, $D_2 = D_3 = D_4 = 4$, $R_1 = 2$, $R_2 = 2.5$, all dimensions in mm [22]. (a) Top view and (b) Bottom view with DGSs. Source: Kumar et al. [22] © [2017] IEEE.

excite left-handed and right-handed CP radiations. The isolation between these ports has been adequately enhanced by introducing multiple DGSs on the ground plane which look like four circular dots and two annular slots. The order of the said improvement in isolation as a function of DGS deployment can be visualized by a systematic study as depicted through Figure 3.20. It is self-explanatory indicating more than 5 dB additional port isolation achieved by the DGS technique [22].

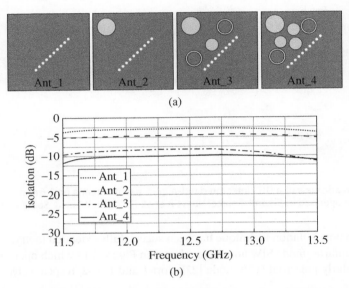

Figure 3.20 Evolution of the antenna geometry described in Figure 3.19 by modification of DGS configuration: (a) the geometries, (b) isolation between the port 1 and port 2 as a function of the said evolution [22]. Source: Kumar et al. [22] © [2017] IEEE.

3.6 Improvement of Antenna Bandwidth

Yet another useful application of DGS is to improve the matching bandwidth of printed antennas. A microstrip patch is actually a resonant structure and as usual of narrowband characteristics. The DGS helps in enhancing their impedance bandwidth by two different ways.

3.6.1 Lowering the Q-Factor

A microstrip patch is equivalent to a resonant cavity with magnetic wall boundary surrounding the edges. Its Q factor actually determines the operating bandwidth which can be enhanced by lowering the Q value. This is a widely known approach since 1970s [23] but the effective use of DGS in controlling Q and increasing the bandwidth has been explored recently [24]. This design was aimed to exploring a new DGS for correcting cross-polarized radiations of a rectangular patch. It used a pair of folded DGSs as shown in Figure 3.21. Parameter g becomes zero when the patch aspect ratio $W/L = 1.6$ and thus half of defect lies beneath the patch. That indeed affects the cavity Q as manifested through its S_{11} furnished in Figure 3.21. The matching bandwidth ($S_{11} < -10$ dB) increases from 850 MHz (with conventional ground plane) to 1040 MHz (with DGS). Similar observation was also reported in [25] with elliptical patches.

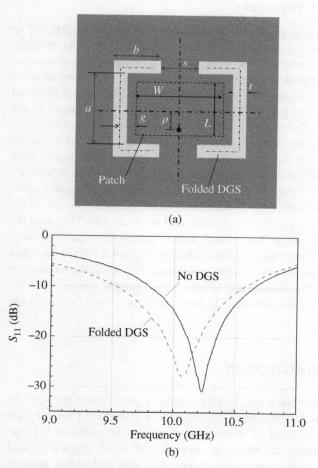

(a)

(b)

Figure 3.21 (a) A probe-fed rectangular microstrip patch integrated with folded DGS, (b) measured S_{11} versus frequency for the patches with and without DGS printed on a 1.57 mm thick RT Duroid 5870 substrate with $\varepsilon_r = 2.33$; parameters: $L = 8.6$, $W = 1.6\,L$, $g = 0$, $\rho = 3.1$, $t = 1.5$, $b = 3.1$, $s = 6$; Unit: mm. Source: Kumar and Guha [24] with permission from The Institution of Engineering and Technology.

A tri band WiMAX/WLAN microstrip antenna [26] simply integrated with dumbbell-shaped DGS under the patch obtained wider impedance bandwidth at all three operating bands. The maximum reported enhancement is 3% at the lowest frequency around 2.45 GHz. Similar benefit was also reported for an L-band design of circularly polarized microstrip antenna by employing Koch curve fractal DGS beneath a circular patch [27]. The antenna geometries in both [26, 27] are shown in Figure 3.22.

3.6.2 Adjusting Higher Resonances

Use of multiple resonances produced by some coupled structures [28, 29] or even by the same radiating element as higher modes [30] are often used as bandwidth widening techniques. The work in [30] actually exploited multiple resonances in a conventional quarter wave monopole along with some adjacent additional resonances produced by coupled resonators. A DGS coupled printed monopole was examined in [31] to realize more than 112% improvement in the impedance bandwidth. The antenna structure is shown in Figure 3.23a. Two interesting engineering have been executed in the feed area. A pair of U-shaped DGSs has been placed beneath the feed line and an open-ended sub has been introduced. Their individual impacts in enhancing the impedance matching are very clearly visible from Figure 3.23b. A signature of the first higher order mode of the printed monopole without DGS is evident around 2.2 GHz. The role of the DGSs is understood from a small study documented in Figure 3.24. They load the monopole capacitively near the higher mode and also drags the input resistance down from 100 to 50 Ω. Eventually, it shifts the higher resonance toward the lower frequency which is adjacent to the primary resonance. Two adjoining resonances thus help in widening the impedance bandwidth. A similar design has been tried later in [32] with a printed monopole demonstrating about 119% matching bandwidth.

3.7 Antenna Miniaturization

A DGS based on its geometry and location tends to perturb the surface current and causes a reactive loading to the associated structure. That property has been strategically utilized in miniaturizing microstrip antennas [33–36]. A capacitive loading shifts the resonance towards the lower frequencies and that effectively reduces the antenna dimension. A design is depicted through Figure 3.25 [35] and the impact of the DGS can be viewed through a set of representative results. A change in the length of a DGS arm from 1 to 5 mm pushes the resonance from 2.6 to 1.75 GHz at the lower band and from 5.4 to 4.4 GHz

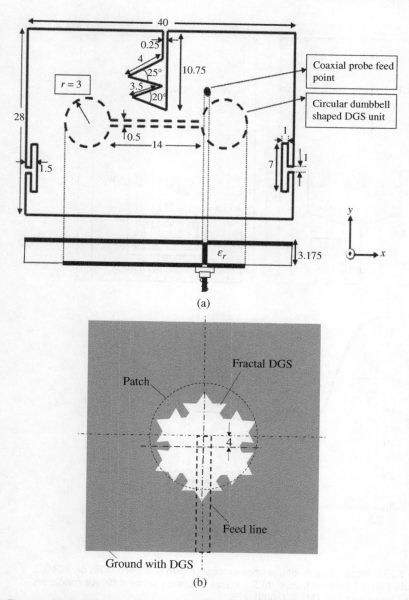

Figure 3.22 Antenna geometries with wider impedance bandwidth achieved by DGS integration technique: (a) probe-fed tri-band rectangular patch with dumbbell shaped DGS [26], (b) Proximity fed circularly polarized circular patch with fractal DGS [27]. Source: (a) Reddy and Vakula [26] © [2015] IEEE, (b) Adapted from Reddy and Vakula [26].

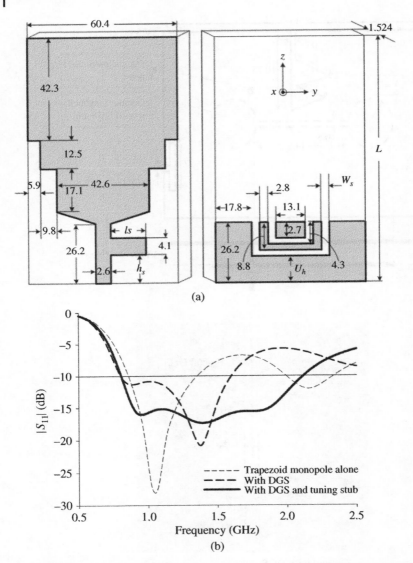

Figure 3.23 Trapezoidal shaped printed monopole with U-shaped DGSs on RO4003 substrate [31]: (a) antenna layout, (b) S_{11} versus frequency under different conditions. Source: Chiang and Tam [31] © [2008] IEEE.

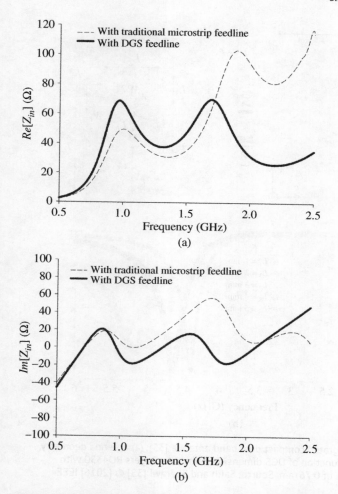

Figure 3.24 A study to examine impact of DGS on the antenna geometry in Figure 3.23a: (a) input resistance as a function of frequency, (b) input reactance as a function of frequency [31]. Source: Chiang and Tam [31] © [2020] IEEE.

at the upper resonance frequency. It claims about 74% miniaturization in patch size. Several other DGS-based antennas such as [26] (Figure 3.22a), [27] (Figure 3.22b), [36] have been benefited in terms of miniaturization of antenna geometries.

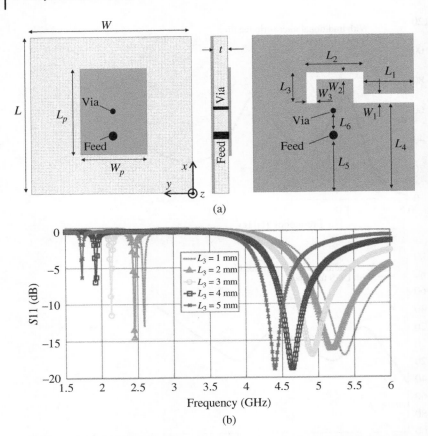

Figure 3.25 DGS integrated compact dual band antenna [35]: (a) antenna geometry, (b) simulated S_{11} as a function of DGS dimension. Substrate Rogers RO4350 with $\varepsilon_r = 3.48$ and thickness of 0.76 mm. Source: Salih and Sharawi [35] © [2016] IEEE.

References

1 V. Radisic, Y. Qian, and T. Itoh, "Broadband power amplifier integrated with slot antenna and novel harmonic tuning structure," *IEEE MTT-S International Microwave Symposium Digest*, pp. 1895–1898, 1998.

2 Y. Horii and M. Tsutsumi, "Harmonic control by photonic bandgap on microstrip patch antenna," *IEEE Microwave Guided Letters,* vol. 9, no. 1, pp. 13–15, 1999.

3 C. S. Kim, J. S. Park, D. Ahn, and J. B. Lim, "A novel 1-D periodic defected ground structure for planar circuits," *IEEE Microwave and Wireless Components Letters,* vol. 10, no. 4, pp. 131–133, 2000.

4 J. S. Lim, C. S. Kim, J. S. Park, D. Ahn and S. Nam, "Design of 10 dB 90° branch line coupler using microstrip line with defected ground structure," *Electronics Letters*, vol. 36, no. 21, pp. 1784–1785, 2000.

5 Y. J. Sung and Y.-S. Kim, "An improved design of microstrip patch antennas using photonic bandgap structure," *IEEE Transactions on Antennas Propagation*, vol. 53, no. 5, pp. 1799–1803, 2005.

6 Image Impedance: Derivation: http://en.wikipedia.org/wiki/Image_impedance.

7 J. P. Thakur and J. Park, "An advance design approach for circular polarization of the microstrip antenna with unbalance DGS feed lines," *IEEE Antennas Wireless Propagation Letters*, vol. 5, pp. 101–103, 2006.

8 Y. Wang, X. X. Yang, G. N. Tan and S. Gao, "Study on millimeter-wave SIW rectenna and array with high conversion efficiency," *IEEE Transactions on Antennas Propagation*, vol. 69, no. 9, pp. 5503–5511, 2021.

9 I. Chang and B. Lee, "Design of defected ground structures for harmonic control of active microstrip antenna," *IEEE Antennas and Propagation Society International Symposium*, vol. 2, pp. 852–855, 2002, San Antonio, TX.

10 D. Ahn, J. S. Park, C. S. Kim, J. Kim, Y. Qian, and T. Itoh, "A design of the low-pass filter using the novel microstrip defected ground structure," *IEEE Transactions on Microwave Theory Techniques*, vol. 49, no. 1, pp. 86–93, 2001

11 H. Liu, Z. Li, X. Sun, and J. Mao, "Harmonic suppression with photonic bandgap and defected ground structure for a microstrip patch antenna," *IEEE Microwave and Wireless Components Letters*, vol. 15, no. 2, pp. 55–56, 2005.

12 Y. J. Sung, M. Kim, and Y.-S. Kim, "Harmonic reduction with defected ground structure of a microstrip patch antenna," *IEEE Antennas Wireless Propagation Letters*, vol. 2, pp. 111–113, 2003.

13 C. Y. D. Sim, M.H. Chang, and B.Y. Chen, "Microstrip-fed ring slot antenna design with wideband harmonic suppression," *IEEE Trans. on Antennas and Propagation*, vol. 62, no. 9, pp. 4828–4832, 2014

14 M. K. Mandal, P. Mondal, S. Sanyal, and A. Chakrabarty, "An improved design of harmonic suppression for microstrip patch antennas," *Microwave and Optical Technology Letters*, vol. 49, no. 1, pp. 103–105, 2007.

15 S. Biswas, D. Guha, and C. Kumar, "Control of higher harmonics and their radiations in microstrip antennas using compact defected ground structures," *IEEE Transaction on Antennas and Propagation*, vol. 61, no. 6, pp. 3349–3353, 2013.

16 S. Biswas, D. Guha, and C. Kumar, "Design of aperture-coupled dielectric resonator antenna free from higher order modes and harmonics," *Microwave and Optical Technology Letters*, vol. 57, no. 8, pp. 1980–1983, 2015.

17 T. L. Wu, Y. M. Pan, and P. F. Hu, "Wideband omnidirectional slotted patch antenna with filtering response," *IEEE Access* vol. 5, pp. 26015–26021, 2017.

18 P. Liu, W. Jiang, W. Hu, S. Y. Sun, and S. X. Gong, "Wideband multimode filtering circular patch antenna," *IEEE Transactions on Antennas and Propagation*, vol. 69, no. 11, pp. 7249–7259, 2021.

19 Y. Chung, S. S. Jeon, D. Ahn, J. I. Choi, and T. Itoh, "High isolation dual polarized patch antenna using integrated defected ground structure," *IEEE Microwave and Wireless Components Letters*, vol. 14, no. 1, pp. 4–6, 2004.

20 Y. Chung, S. S. Jeon, S. Kim, D. Ahn, J. I. Choi, and T. Itoh, "Multifunctional microstrip transmission lines integrated with defected ground structure for RF front-end application," *IEEE Transactions on Microwave Theory Techniques*, vol. 52, no. 5, pp. 1425–1432, 2004.

21 A. H. Hussein, H. H. Abdullah, M. A. Attia and A. M. Abada, "S-band compact microstrip full-duplex Tx/Rx patch antenna with high isolation," *IEEE Antennas and Wireless Propagation Letters*, vol. 18, no. 10, pp. 2090–2094, 2019.

22 K. Kumar, S. Dwari, and M. K. Mandal, "Broadband dual circularly polarized substrate integrated waveguide antenna," *IEEE Antennas and Wireless Propagation Letters*, vol. 16, pp. 2971–2974, 2017.

23 K. Carver and J. Mink, "Microstrip antenna technology," *IEEE Transactions on Antennas and Propagation,* vol. 29, no. 1, pp. 2–24, 1981, DOI: 10.1109/TAP.1981.1142523.

24 C. Kumar and D. Guha, "Defected ground structure (DGS)-integrated rectangular microstrip patch for improved polarization purity with wide impedance bandwidth," *IET Microwaves, Antennas & Propagation*, vol. 8, no. 8, pp. 589–596, 2014.

25 C. Kumar and D. Guha, "Linearly polarized elliptical microstrip antenna with improved polarization purity and bandwidth characteristics," *Microwave and Optical Technology Letters*, vol. 54, no. 10, pp. 2309–2314, 2012.

26 B. R. S. Reddy and D. Vakula, "Compact zigzag-shaped-slit microstrip antenna with circular defected ground structure for wireless applications," *IEEE Antennas and Wireless Propagation Letters*, vol. 14, pp. 678–681, 2015.

27 P. R. Prajapati, G. G. K. Murthy, A. Patnaik, and M. V. Kartikeyan, "Design and testing of a compact circularly polarised microstrip antenna with fractal defected ground structure for L-band applications'" *IET Microwaves, Antennas & Propagation*, vol. 9, no. 11, pp. 1179–1185, 2015.

28 K. F. Lee, K. M. Luk, K. F. Tong, and S. M. Shum, "Experimental and simulation studies of the coaxially fed U-slot rectangular patch antenna," *IEE Proceedings – Microwave, Antennas and Propagation*, vol. 144, no. 5, pp. 354–358, 1997.

29 D. Guha, Y. M. M. Antar, A. Ittiboon, A. Petosa, and D. Lee "Improved design guidelines for the ultra wideband monopole-dielectric resonator antenna," *IEEE Antennas and Wireless Propagation Letters,* vol. 5, pp. 373–376, 2006.

30 D. Guha, D. Ganguly, S. George, C. Kumar, M. T. Sebastian, and Y. Antar, "New design approach for hybrid monopole to achieve increased ultra-wide bandwidth," *IEEE Antennas and Propagation Magazine*, vol. 59, no. 1, pp. 139–144, 2017.

31 K. H. Chiang and K. W. Tam, "Microstrip monopole antenna with enhanced bandwidth using defected ground structure," *IEEE Antennas and Wireless Propagation Letters*, vol. 7, pp. 532–535, 2008.

32 H. F. Abutarboush, W. Li, and A. Shamim "Flexible-screen-printed antenna with enhanced bandwidth by employing defected ground structure," *IEEE Antennas and Wireless Propagation Letters*, vol. 19, no. 10, pp. 1803–1807, 2020.

33 M. Taouzari, J. El Aoufi, A. Mouhsen, H. Nasraoui, and O. El Mrabat, "900 MHz and 2.45 GHz compact dual-band circularly-polarized patch antenna for RFID application," *Proceedings of Conference on Microwave Techniques*, 2015, pp. 1–4, Pardubice, Czech Republic, DOI: 10.1109/COMITE.2015.7120226.

34 S. K. Singh, P. Consul, and K. K. Sharma, "Dual-band gap coupled microstrip antenna using L-slot DGS for wireless applications," *Proceedings of International Conference on Computing, Communicating & Automation*, 2015, pp. 1381–1384, Greater Noida, India, DOI: 10.1109/CCAA.2015.7148595.

35 A. A. Salih and M. S. Sharawi, "A dual band highly miniaturized patch antenna," *IEEE Antennas and Wireless Propagation Letters*, vol. 15, pp. 1783–1786, 2016.

36 R. S. Ghoname, M. A. Mohamed, and A. E. Hennawy, "Reconfigurable compact spider microstrip antenna with new defected ground structure," *International Journal of Computer Applications*, vol. 54, no. 5, pp. 47–52, 2012.

4

DGS to Control Orthogonal Modes in a Microstrip Patch for Cross-Pol Reduction

4.1 Introduction

Different aspects of DGS have been discussed in Chapters 1–3 which are mostly associated with microwave printed circuits and components. Its other application to microstrip antennas has been addressed here. Control of unwanted modes underneath microstrip patches is the area of its application which does not necessarily demand a DGS to be resonant always. In reality, a microstrip patch is always subjected to a limited number of weakly coupled unwanted modes that cause cross-polarized (XP) radiation. Such unwanted radiation has been a serious concern to the antenna engineers. The discussion, therefore, begins with a quick recap of the resonant modes of our interest in two most commonly used microstrip geometries which follows the principle of weakening the unwanted modes by DGS-based engineering. A wide range of geometries, their natural evolution and the operating principles have been portrayed.

4.2 Understanding of Radiating Modes in Microstrip Patches

A microstrip patch antenna is a thin metallic section printed or placed on a grounded substrate. The substrate can have a nominal range of relative permittivity (ε_r) upto 10 starting from 1 (when it is simple air). The patch creates a cavity underneath it bound by two electric walls on the top (the patch itself) and bottom sides (the ground plane), and magnetic boundaries around the edges. Figure 4.1 helps in understanding the same. The Radio Frequency (RF) energy is coupled to this cavity using different types of feeds like a vertically protruded coaxial probe, printed coplanar microstrip line, or centrally located aperture on the ground plane. At a given frequency, this energy gets distributed within the cavity depending on its shape, size, boundary conditions, and finally, the matching with

Defected Ground Structure (DGS) Based Antennas: Design Physics, Engineering, and Applications,
First Edition. Debatosh Guha, Chandrakanta Kumar, and Sujoy Biswas.
© 2023 The Institute of Electrical and Electronics Engineers, Inc. Published 2023 by John Wiley & Sons, Inc.

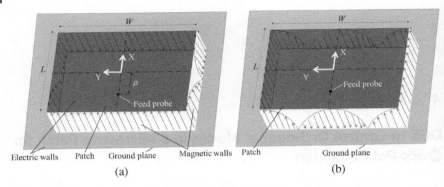

Figure 4.1 Modal distribution of electric fields around the patch boundary in a rectangular patch: (a) dominant TM_{10} mode; (b) higher order TM_{02} mode. Source: Balanis [2] John Wiley & Sons.

the feed impedance. The phenomenon is called "resonance." The resulting spatial distributions of the electric and magnetic fields inside the cavity are designated as "modes" or "resonant modes." Because of the magnetic wall boundary around a patch, the electric fields tend to fringe out at the edges. The fringing fields indeed create radiating apertures causing electromagnetic radiations to the free space. An idea can be obtained for patches with regular shapes where analytical technique or simplified models can be easily developed [1, Chapter 2]. Two most commonly used geometries are discussed below.

4.2.1 Rectangular Patch

Let us consider a rectangular patch excited by a probe located on the x-axis as shown in Figure 4.1. The probe location is asymmetrical with respect to the y-axis. Thus, resonances should naturally set in along x-axis with E-fields within the cavity polarized to z and associated H-fields transverse to z. Therefore, the modes are called "Transverse Magnetic to z" (TM^z) or simply TM_{mn}. Subscripts m and n indicate the number of field variations along x and y axes, respectively where each half-cycle or $\lambda/2$ variation counts "1." For obtaining a resonance with the lowest possible frequency f_o, the electric field distribution looks like that in Figure 4.1a which represents TM_{10} mode. In here, E-fields show one half-cycle variation along x ($m = 1$), and absolutely no variation along y ($n = 0$) [2, 3]. This TM_{10} mode is responsible for the broadside radiations from a rectangular patch. The nature of the E-fields appearing on its four edges is also visible in Figure 4.1a. The edges bearing unipolar E across the patch-width are termed as "radiating edges" through which the fields fringe out. The rest two edges are ideally "non-radiating" as the resultant E across each of those edges has to be ideally zero. The fringing fields for TM_{10} mode eventually result in liner x-polarized E as the radiating source and causes radiation surrounding the patch with its peak along the z-axis. The radiated E fields remain aligned to x, the axis of TM_{10} resonance, and hence they are commonly termed as "co-polarized (CoP) radiations."

From the sketch in Figure 4.1a, one can conclude that a rectangular patch resonating with TM_{10} mode radiates purely co-polarized fields oriented towards **x**. But, in reality, it is not the absolute truth. The feeding structure causes some kind of perturbation inside the microstrip cavity and generates weakly coupled higher order modes. Under a rectangular microstrip, TM_{0n} become the most probable ones which are orthogonal to E_{TM10} and result in cross-polarized (XP) fields. The historical investigation dates back to 1983 [3–5]. The scientific definition could be found in [6]. The presence of those weakly generated byproducts creates an aberration and that is manifested through radiation fields. A set of representative practical radiation patterns from a TM_{10} mode driven rectangular patch is shown in Figure 4.2. The measured XP radiation bears a signature of TM_{0n} mode, especially on the H-plane (Figure 4.2a), where the value of n has to be $n = 2$ and this is true for all rectangular geometries. If we consider TM_{02} to be present exclusively beneath the patch, its ideal field orientation would look like that in Figure 4.1b.

This clearly indicates that there is absolutely no scope of creating x-polarized E-fields out of TM_{02} mode, although enough y-polarized fields appear across the narrow edges of the patch ($y = \pm W/2$). Those y-polarized fields in Figure 4.1b are unipolar and their fringe out components would produce a resultant $E_y = 0$ over xz-plane for $y = 0$. This means that the XP level over the E-plane should be ideally zero. This theoretical estimation is endorsed by the measured data as furnished through Figure 4.2b. They are more than 35 dB below the peak radiation. But the individual edge-fields (Figure 4.1b) contribute to the H-plane XP radiation (Figure 4.2a) with peaks nearing 45°–50° and a null at boresight.

As mentioned before, the higher order mode is weakly coupled and as a result, causes radiation of relatively weaker intensity. But it increases with the increase in patch aspect ratio W/L [7]. A typical study is shown in Figure 4.3 [7]. Some XP related results furnished in [8] are frequently referred to as evidences for rectangular microstrip patches, e.g. in [1; Figure 4.17]. This may mislead one since the design in [8] bears offset-fed rectangular patches.

Ideally, the radiation patterns should be symmetrical with reference to the boresight. This is observed to be perfect on the H-plane, but not on the E-plane. The asymmetry in the feed-location is the primary reason behind this. It gets considerably reduced for a symmetrical feed like aperture-coupling. A more exhaustive study on the higher order modes in a rectangular patch and their radiation patterns is available in [9].

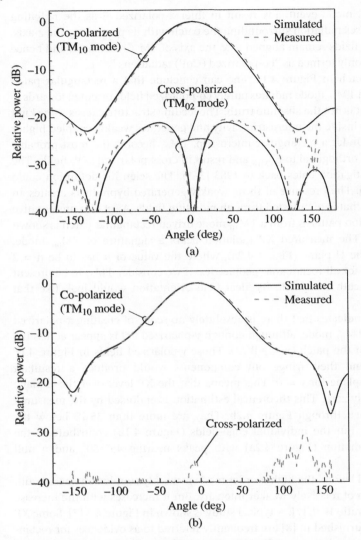

Figure 4.2 Typical measured and simulated radiation patterns of a rectangular patch designed at 6.3 GHz with $W/L = 1.25$, $L = 14.2$ mm, $\rho = 3$ mm and printed on a substrate with thickness 1.575 mm and $\varepsilon_r = 2.33$: (a) H-plane; (b) E-plane.

4.2.2 Circular Patch

The nature of the modal fields beneath a circular patch is a bit different compared to those under rectangular geometry. Here the fields vary as a function of the radial distance ρ from patch center along with the azimuthal angle φ [10]. TM$_{11}$

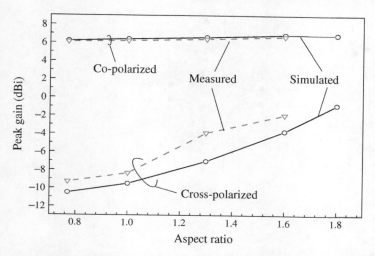

Figure 4.3 Variation of peak gains for CoP and XP fields in $\varphi = 90°$ plane with varying W/L of a rectangular patch printed on a 1.575 mm thick substrate with $\varepsilon_r = 2.33$ and resonating near 10 GHz. Parameters as in [7; Table 1].

is the primary radiating mode as sketched in Figure 4.4a. The fringing electric fields along with the surface conduction current are linearly polarized along the x-axis and they cause linearly polarized (LP) radiations. A pure TM_{11} mode ideally cannot produce any cross-polarized radiation which requires some degree of y-polarized E-fields. But a practical circular patch produces considerable cross-polarized radiations as in rectangular geometries. Again the reason is the presence of an unwanted higher order mode [10].

Weakly generated TM_{21} mode has been identified as the cause whose E-field orientations are shown in Figure 4.4b. Here also, the resulting fringing fields over the xz-plane ($y = 0$) are mutually out of phase and hence contribute almost no radiation along bore sight. This is reflected in the E-plane patterns. But across the H-plane (yz-plane), the y-polarized TM_{21} fields cause oblique XP radiation with broadside null. They can be visualized from a set of representative radiation patterns as shown in Figure 4.5. The measured XP level in Figure 4.5a is below −35 dB and ensures negligibly small XP radiations over the entire E-plane. Figure 4.5b, in contrast, indicates relatively high XP level over the H-plane peaking around 45°–50°. But there may be some variations.

Sometimes a circular patch may exhibit considerably high XP values even over the E-plane. An example is shown in Figure 4.6a where the XP pattern follows the co-polarized radiation by TM_{11} mode, but with a weaker intensity. This case was first addressed in [11] and subsequently investigated in [12, 13]. According to them, orthogonally polarized TM_{11} mode is an additional factor behind such XP

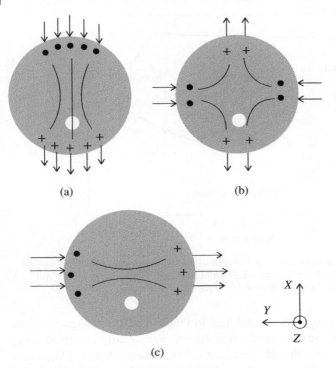

(a)   (b)

(c)

Figure 4.4 Field and surface current distributions in a circular microstrip patch for different modes. Symbols "+" and "•" indicate electric fields between patch and ground, fringe fields are shown outside the patch boundary with arrows, surface currents are represented by solid lines, white dot indicates the feed location: (a) TM_{11} mode or dominant mode; (b) TM_{21} mode; (c) Orthogonal component of the dominant TM_{11} mode (OCDM). Source: Kumar and Guha [13] © IEEE.

behavior and it is directly correlated to the order of matching the input impedance of the feed. It looks like Figure 4.4c and is called "orthogonal component of dominant mode" (OCDM) [11].

When the feed is perfectly matched, TM_{11} mode gets excited maximally along with a fraction of input energy coupled to TM_{21} mode. The matching bandwidth becomes narrow and it appears as the best suited condition for minimal XP fields. But in practice, we require sufficient operating bandwidth and design tolerance.

This is achievable when the impedance matching is relatively weak. Marginal change in feed location is the clue to achieve the same and in such cases, a small fraction of the input signal may get coupled to OCDM in addition to TM_{21} mode. Thus, the signature of both OCDM and TM_{21} are present in Figure 4.6a,b.

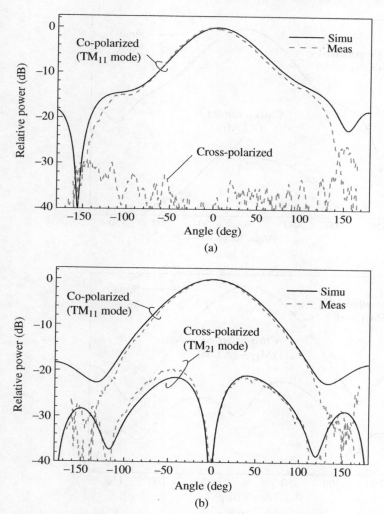

Figure 4.5 Radiation patterns of a circular patch resonating around 5.93 GHz: (a) E-plane; (b) H-plane [13]. Patch diameter = 18 mm, probe location $\rho = 2.8$ mm, substrate thickness = 1.575 mm with $\varepsilon_r = 2.33$. Source: Kumar and Guha [13] © IEEE.

An asymmetry in H-plane XP pattern with 1–2 dB difference between the peaks is evident from Figure 4.6b and the reason is the presence OCDM. The OCDM-driven fringing fields appear in phase with those due to TM_{21} at one edge and out of phase at the other edge over the H-plane. ***Thus, TM_{21} and OCDM are of particular interests in dealing with XP issues in a circular patch.***

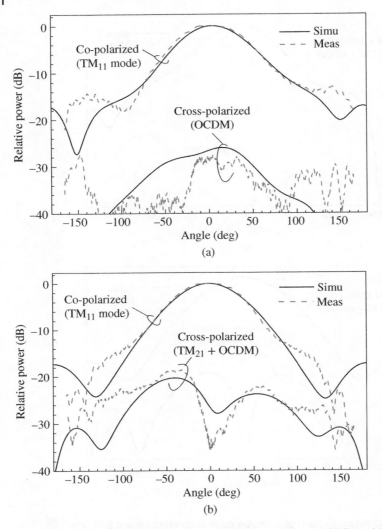

Figure 4.6 Radiation patterns of a circular patch operating at 5.87 GHz with probe location $\rho = 2.25$ mm. Other parameters as in Figure 4.5: (a) E-plane; (b) H-plane. Source: Kumar and Guha [13] © IEEE.

4.3 What Were the Known Methods to Deal with the Cross-Polarized Fields?

As discussed earlier, the cross-polarized fields are not the desired output from a linearly polarized antenna. This limits its applications and hence the antenna researchers have been trying to explore several techniques to improve the polarization purity by controlling or suppressing this unwanted XP radiation. A few of them are referred to here which use band-gap techniques [14], differential or balanced feeding [15-17], ground plane shaping [18, 19], shortening by vias [20, 21], etc. A few typical geometries are shown in Figure 4.7.

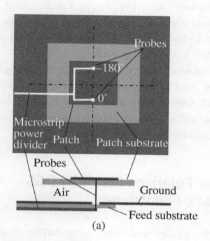

(a)

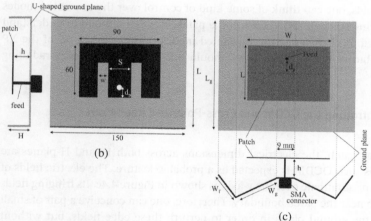

(b)   (c)

Figure 4.7 Different techniques for suppressing XP fields in microstrip antenna designs: (a) use of dual feed. Source: Petosa et al. [15] With permission of Institution of Engineering and Technology; (b) U-shaped ground plane. Source: Hsu and Wong [18] © IEEE; (c) W-shaped ground plane. Source: Wong et al. [19] © IEEE.

The balanced phasing by dual probe is one effective method revealing H-plane XP suppression by 10–15 dB [15], but at the cost of using an additional feed network. Other techniques [18, 19] convert the flat planar ground plane to shaped 3D bulky structures.

A technique has been shown in [20], where small pins are used as perturbations at some strategic locations on the radiating patch surface. This is primarily aimed to achieving flexibility in matching resonance. But in addition to this, the pin combinations show reduction in XP level by controlling modal surface current. Another study [21] has explored shorting pins in a rectangular patch to discriminate the target higher mode and hence improve the XP isolation. More recent investigations have used spike loading on the patch surface [22] or on the ground plane [23] for weakening the higher order mode.

All these techniques are able to reducing the XP radiations over the H-plane and limiting within about 10 dB. They lead to increased complexity in fabrications, volume, weight, and cost. Substrates other than air may not be suitable in most of techniques. Thus, they lose the advantages of planar low profile feature and integrability with the associated circuits sharing the common platform.

4.4 Suppression of Cross-Polarized Fields by DGS Integration Technique: Coax-Fed Patches

With the above knowledge of the modes and their contribution towards the radiation fields, one can think of some kind of control over the unwanted modes by introducing strategic DGS beneath the patch. It is not difficult to understand that we need to disintegrate the unwanted modes in order to get rid of the XP radiations, but at the same time, we should not disturb the primary radiating mode.

4.4.1 Controlling the OCDM and Cross-Polarized Radiations in E-Plane

In a circular patch, the electrical dimensions across both E- and H-planes are same and as such OCDM is expected as a probable feature. The electric fields of the OCDM predominate over the y-axis as shown in Figure 4.4c. Its fringing fields concentrate near the patch boundary. Therefore, one can conceive a pair of small defects on the ground plane in order to perturb these edge fields, but without disturbing the primary mode of resonance. The scheme is shown in Figure 4.8 and a complete disappearance of the OCDM fields can be visualized in Figure 4.9. This indeed was the first application of DGS to controlling the antenna radiations and the result was verified experimentally with two sets of prototypes operating at 3.6 GHz [11], and 5.9 GHz [13]. The diameter of the dot-shaped DGS was close to $0.07\lambda_0$ and their centers were aligning with the patch boundary. About 5 dB

Figure 4.8 Schematic diagram of a circular microstrip integrated with "dot-DGS": (a) view from the ground plane side; (b) cross sectional view [11, 13]. Source: Kumar and Guha [13] © IEEE.

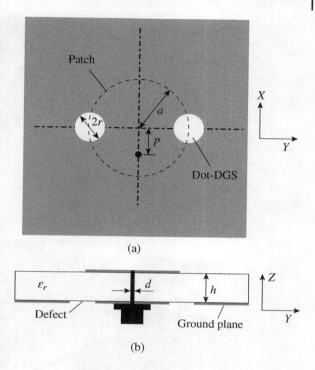

(a)

(b)

suppression in the XP field without affecting the co-polarized radiation was achieved [11] as shown in Figure 4.10.

Now the question is: will such dot-shaped DGS be equally efficient in interacting with TM_{21} mode? This TM_{21} mode is of more general interest since this is the main player in generating XP fields, especially affecting the H-plane radiations. This has been addressed in the subsequent sections.

4.4.2 Controlling of TM_{21} Mode and Cross-Polarized Radiations in Circular and Elliptical Patches

It is interesting to note that, the real issue of significant cross-polar fields over the H-plane cannot be addressed by small dot-shaped defects. Indeed, TM_{21} mode is unlike OCDM and its fringing fields are distributed over the patch boundary. For an effective interaction with the fringing fields, the DGS is supposed to follow the patch contour. Thus, a ring-shaped DGS, as shown in Figure 4.11a,

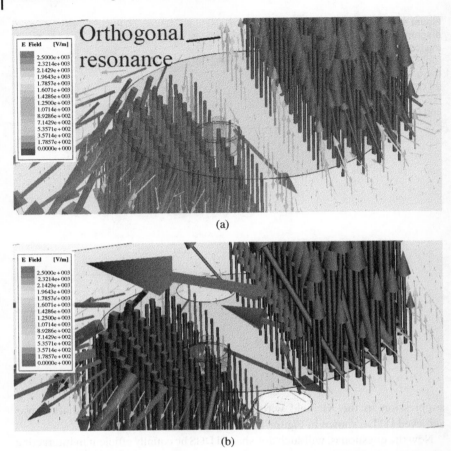

(a)

(b)

Figure 4.9 Simulated E-field obtained at resonance with and without DGS beneath a circular microstrip patch [11]: (a) without DGS: OCDM present; (b) with dot-DGS: OCDM eliminated. Parameters: $a = 15$ mm, $\rho = 5$ mm, $r = 3$ mm, $d = 0.76$ mm, $h = 1.575$ mm, $\varepsilon_r = 2.32$, ground plane: $0.7\,\lambda_0 \times 0.7\,\lambda_0$. Source: Guha et al. [11] © IEEE.

was conceived as a useful solution [25]. An optimum value of the width t is about 0.04–0.05 λ_0 for a substrate having thickness 1.575 mm and $\varepsilon_r = 2.33$. The measured radiation properties using a ring-DGS are shown in Figure 4.12. An improvement by 4–5 dB is evident. But, can we improve it further? No, ring-DGS cannot be brought too close to the patch as it would disturb the primary radiating TM_{11} mode [13].

In order to overcome the said limitation of a "ring," it was truncated to give a shape of *arc* as shown in Figure 4.11b [13]. The aim is to leave the main fringing

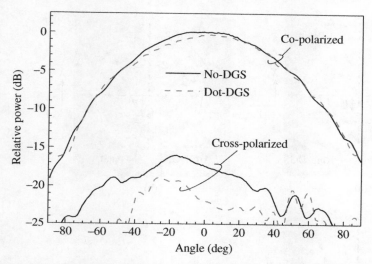

Figure 4.10 Measured E-plane radiation patterns ($f = 3.6$ GHz) of the antennas examined in Figure 4.9. Parameters as in Figure 4.9. Source: Guha et al. [11] © IEEE.

fields over the E-plane unaffected. This "arc-DGS" now can be brought very close to the patch without disturbing the primary radiating fields. Its width t remains unchanged. The angular span of the arc is found to be frequency independent as the studies both C- and X-bands [13, 25, 26] reveal $2\alpha \approx 130°$.

The measured XP characteristics using *arc* DGS are shown in Figures 4.13 and 4.14. The E-plane radiations shown in Figure 4.13a hardly reflect any effect due to DGS. Whereas, the H-plane data obtained near the mid-band frequency are shown in Figure 4.13b. They indicate 9–10 dB reduction in the XP level. The order of the reduction has been examined in Figure 4.14 for the entire operating band. The arc-DGS is found to influence neither the CoP radiations nor the input impedances as examined in Figures 4.13–4.15.

An *ellipse* is a more general form than a *circle* and hence, a question may arise about the role of arc-shaped DGS if the patch geometry turns to an ellipse. A schematic view is shown in Figure 4.11c which was studied in X-band [13, 27]. The slot width is not sensitive to major to minor axis ratio (a/b). One set of representative data [13] is shown in Figure 4.16. It was observed that the DGS becomes more effective for larger b/a resulting in higher suppression in XP radiation [27]. Indeed, in an elliptical patch with larger b/a, the DGS comes closer to the patch. This in turn reduces the Q-value of the resonant structure and helps in increasing the impedance bandwidth, up to about 2% for $b/a = 1.6$ [27].

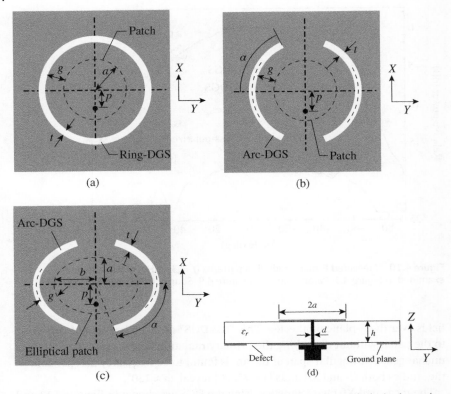

Figure 4.11 Schematic diagrams of DGSs employed to circular and elliptical microstrip patches for suppressing TM_{21} mode: (a) ring-DGS; (b) arc-DGS with circular patch; (c) arc-DGS with elliptical patch; (d) cross sectional view of the DGS integrated patch. Source: Kumar and Guha [13] © IEEE.

4.4.3 Controlling TM_{02} Mode in a Rectangular Patch and H-Plane Cross-Polarized Radiations

The field distributions in Figure 4.1 indicate that the unipolar fringing fields due to TM_{02} mode which are responsible for the XP generation are primarily aligned over the H-plane edges. This gives a clue of deploying DGSs beneath those edges. Then the DGS automatically takes a shape like "[" and "]" as shown in Figure 4.17. They may be called as "folded-DGS" [7, 28]. Some interesting information may be noted based on an investigation carried out around 10 GHz [7]:

i. The width of the DGS remains the same so long the operating frequency remains unchanged.
ii. The optimum width of the DGS depends only on the frequency and substrate properties.

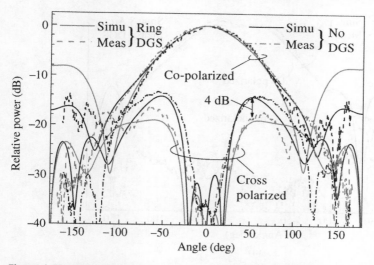

Figure 4.12 Measured and simulated H-plane radiation patterns of circular patches with and without ring-DGS ($f = 10.05$ GHz). Parameters: $\varepsilon_r = 2.33$, $h = 1.575$ mm, $a = 5$ mm, $d = 1.3$ mm, $\rho = 1.67$ mm, $g = 2.25$ mm, $t = 1.5$ mm, ground plane: 60 mm × 60 mm. Source: Kumar and Guha [13] © IEEE.

iii. Optimization of the parameter "s" needs a special care since the same (below a certain value) may degrade the primary radiation.
iv. Other parameters need to be optimized on the basis of maximum suppression of the XP level.

Choice of the aspect ratio W/L is an important factor for a rectangular patch. Figure 4.18 depicts the XP properties of one such X-band patch with $W/L = 1.3$ [7]. As mentioned earlier, a DGS should not affect the co-polar radiation in any case. The principal plane radiations shown in Figure 4.18 also endorse the same. Only the H-plane XP radiations are found to be suppressed by about 10 dB. This DGS configuration effectively performs well for as possible categories like $W/L > 1$, $W/L = 1$, and $W/L < 1$ [7]. The parameter s is not a function of W/L, but it depends on the frequency of operation and the substrate parameters only.

Another interesting observation is noted which looks very similar to the case of *ellipse*. Like larger b/a ratio, larger W/L drags the optimum DGS closer to patch edges. Even it may enter underneath the patch for $W/L = 1.6$ [7]. This actually lowers the Q value and results in an increase in matching bandwidth by about 2%.

4.4.4 Visualization of the Modal Fields and the Effect of the DGSs

The electromagnetic simulation tools such as [24, 29, 30]. enable one in visualizing the dynamics of the fields and currents in an antenna body and that pictorial representation helps to understand how they get influenced in presence of a DGS.

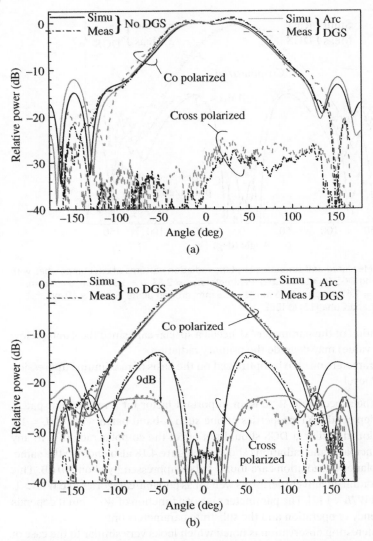

Figure 4.13 Measured and simulated radiation patterns of circular patches with and without arc-DGS ($f = 10.05$ GHz): (a) E-plane; (b) H-plane. Parameters as in Figure 4.12 with $\alpha = 65.5°$. Source: Kumar and Guha [13] © IEEE.

Figure 4.19 portrays some representative simulated results for both rectangular and circular patches. The distribution of substrate field appears interesting as both of them indicate an asymmetry with reference to y-axis. One half gets stronger fields than the other half. The reason is the asymmetry in the resultant fringing fields around the patch edges.

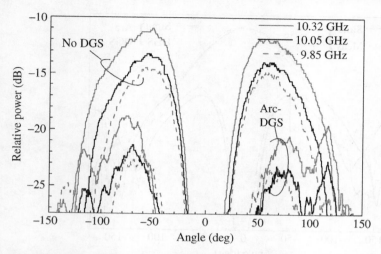

Figure 4.14 Measured cross-polarized radiation patterns in H-plane over the matching bandwidth ($S_{11} \leq -10$ dB) of circular patches with and without arc-DGS (corresponding co-pol peak normalized to 0 dB). Parameter as in Figure 4.13. Source: Kumar and Guha [13] © IEEE.

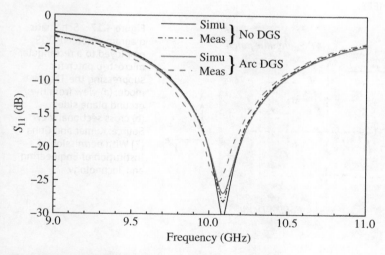

Figure 4.15 Reflection coefficient versus frequency of probe-fed circular patches with and without arc-DGS. Parameter as in Figure 4.13. Source: Kumar and Guha [13] © IEEE.

Figure 4.20 helps in analyzing them. Deep black and grey shades have been chosen to represent the dominant and higher modes, respectively under each patch [31]. The fields due to the dominant modes (deep black) have been placed outside the patch boundary and those due to the higher order mode (light gray) inside the

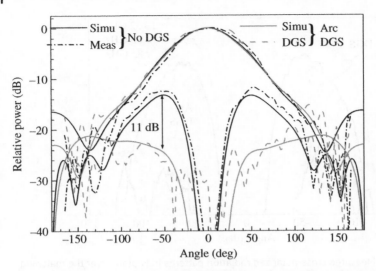

Figure 4.16 H-plane radiation patterns of elliptical patches ($b/a = 1.3$) with and without arc-DGS ($f = 10$ GHz). Parameters: $a = 5$ mm, $b = 6.5$ mm, $\rho = 2.2$ mm, $t = 1.5$ mm, $\alpha = 59°$, $g = 1$ mm, $h = 1.575$ mm, $\varepsilon_r = 2.33$ ground plane 60 mm × 60 mm. Source: Kumar and Guha [13] © IEEE.

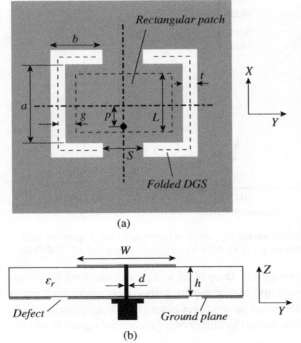

(a)

(b)

Figure 4.17 Schematic diagram of folded-DGS employed to a rectangular microstrip patch for suppressing the TM_{02} mode: (a) view from the ground plane side; (b) cross sectional view. Source: Kumar and Guha [7] With permission of Institution of Engineering and Technology.

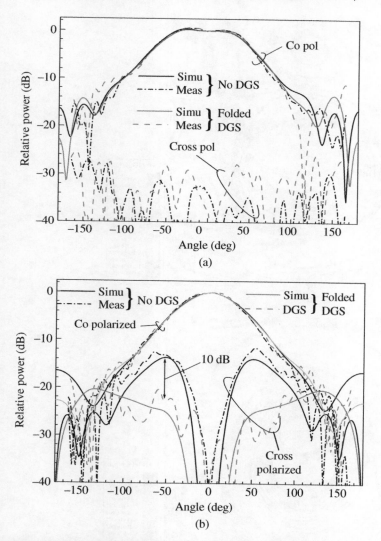

Figure 4.18 Radiation patterns of rectangular patches ($W/L = 1.3$) with and without folded-DGS ($f = 10.15$ GHz): (a) E-plane; (b) H-plane. Parameters: $h = 1.575$ mm, $\varepsilon_r = 2.33$, $W = 11.05$ mm, $L = 8.5$ mm, $\rho = 2.2$ mm, $t = 1.5$ mm, $g = 0.75$ mm, $s = 6$ mm; ground plane 60 mm × 60 mm. Source: Kumar and Guha [7] With permission of Institution of Engineering and Technology.

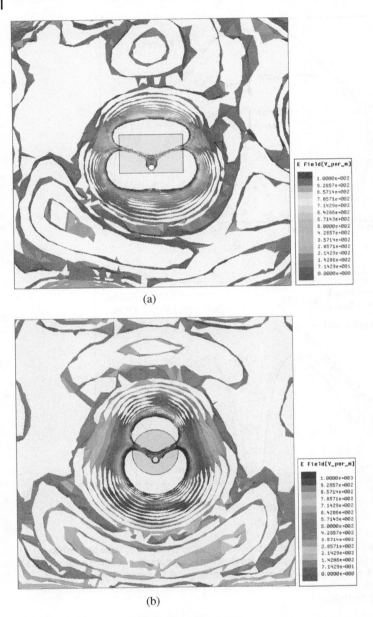

Figure 4.19 Simulated substrate fields obtained near 10 GHz using conventional ground plane: (a) rectangular patch with $L = 8.6$ mm, $W/L = 1.6$, $\rho = 3.1$ mm; (b) circular patch with radius $= 5$ mm, $\rho = 1.67$ mm. Substrate: $\varepsilon_r = 2.33$, thickness $= 1.575$ mm, ground plane 60 mm $\times$ 60 mm.

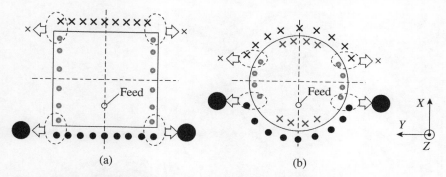

Figure 4.20 Schematic diagram indicating Z-polarized electric fields with symbols: Deep black means dominant mode (pictorially placed outside the patch boundary) and gray shade means higher order mode (pictorially placed inside the patch boundary). (a) square patch with TM_{10} and TM_{02} modes; (b) circular patch with TM_{11} and TM_{21} modes. Source: Kumar and Guha [31] © IEEE.

boundary. The higher mode produces unipolar edge fields and interacts with the dominant mode near the mutually overlapped or corner regions. In Figure 4.20, the higher mode fields are out of phase with respect to the dominant mode in the upper half section. It is of an opposite nature in the lower half section of the patches. The resultant has been pictorially shown by small "crosses" and large "dots" and they symbolically represent asymmetry in the field intensity as an overall outcome. They also logically corroborate the asymmetry reflected through the simulated field portrays in Figure 4.19.

As already conjectured, the purpose of using DGS is to weaken the higher order mode. If it is true, then the asymmetry in the substrate fields as observed in Figure 4.19 should disappear in presence of DGS. The reality is evidenced in Figure 4.21. The expected field symmetry due to the dominant mode resonance, which is lost in conventional coax-fed patches (Figure 4.19) has been regained in Figure 4.21 by introducing appropriate DGS.

4.4.5 Universal DGS: Applicable to Both Circular and Rectangular Patch Geometries

The arc- and the folded-DGS are quite successful in improving the cross-polarized isolation of the circular/elliptical and rectangular microstrips, respectively. Consistent improvement ranging from 9 to 15 dB over the band was achieved [7, 13]. Here, these DGS shapes need to follow the patch boundary and therefore, it demands patch specific precise design every time. Additionally, the length of the DGS also gets restricted as it has to stay out of the fringing field territory of the radiating mode. Such restrictions also limit the performance in terms of the order of XP suppression.

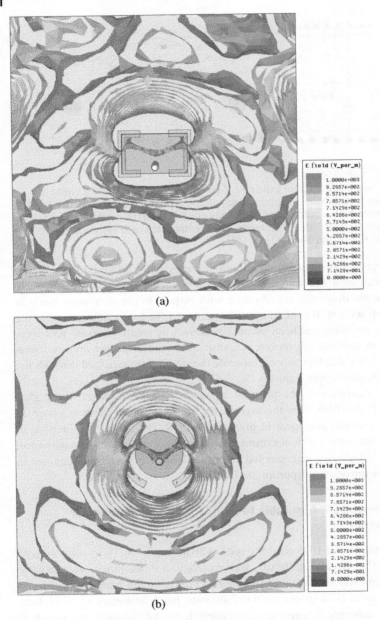

Figure 4.21 Simulated substrate fields obtained near 10 GHz using DGS integrated ground plane: (a) rectangular patch as in Figure 4.19.a with DGS parameters $t = 1.5$ mm, $g = 0$ mm, $a = 7.1$ mm, $b = 4.6$ mm, $s = 6$ mm; (b) circular patch as in Figure 4.19.b with DGS parameters $g = 2.25$ mm, $t = 1.5$ mm, $\alpha = 65.5°$.

To overcome these limitations, a geometry independent DGS was conceived [31]. A pair of linear DGSs was applied along the H-plane edges of circular or rectangular patch as shown in Figure 4.22. It is easy to understand here that an engineer is free to design a DGS as a linear slot of length l. In fact, $l \approx \lambda_0$, λ_0 being the free space wavelength corresponding to the mid-band frequency. The fringing electric fields around the non-radiating edges get trapped in the resonant slot. The nature of the slot fields ideally looks like Figure 4.22c. The full wave variation within the slot results in zero equivalent fields and makes the slot non-radiating. Their performances are clearly evident through a set of representative results shown in Figure 4.23 [31]. Significant XP suppression by 16 dB in a rectangular patch and about 11 dB in a circular patch is achieved. This low XP feature is spanning over a wide angular range around the boresight.

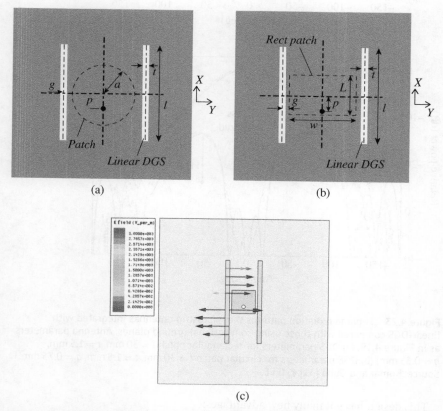

(a)

(b)

(c)

Figure 4.22 Linear DGS integrated microstrip patch: (a) circular patch; (b) rectangular patch; (c) electric field trapped in the slots (one slot is portrayed at a time due to clarity of fields). Source: Kumar and Guha [31] © IEEE.

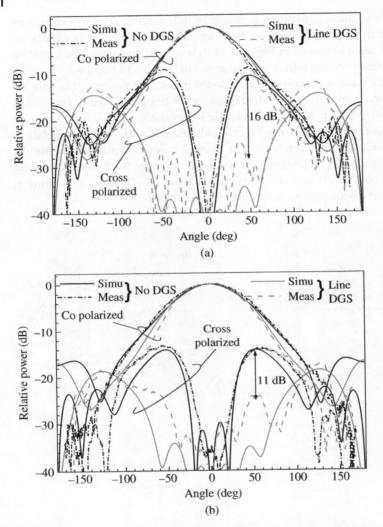

Figure 4.23 H-plane radiation patterns the microstrip antennas integrated with linear-DGS compared with those using conventional ground planes. Antenna parameters as in Figure 4.19: (a) DGS parameters for rectangular patch $l = 30$ mm, $t = 1.5$ mm, $g = 0.35$ mm; (b) DGS parameters for circular patch $l = 30$ mm, $t = 1.5$ mm, $g = 0.75$ mm. Source: Kumar and Guha [31] © IEEE.

This design has got many new advantages:

i. **Simplified design parameters**: The length of the defect is always one full wavelength ($l = \lambda_0$). The proximity parameter g needs to be determined on the basis of the optimum coupling with the fringing fields. This resonant DGS imposes only one requirement in the ground plane dimension (to be $>l$ along the axis of resonance) in order to accommodate the linear DGS. The width of the defect is typically $t \approx 0.04–0.05\ \lambda_0$ and is not a widely varying parameter.

ii. **Geometry independent design:** The radiator or patch geometry is not the concern while designing a resonant DGS. This is a major freedom to a designer [31].

iii. **Superior performance:** This DGS is found to be more effective compared to other non-resonant-type defects. This is revealed through the order of XP suppression when compared with those due to arc-DGS [13] or folded-DGS [7].

4.4.6 DGS for Triangular Microstrip Patch

Triangle is yet another shape of microstrip patches and bears some special features. It is not as common as rectangle or circle in terms of microstrip antenna and arrays. Typically, it is of very high Q and hence narrow matching bandwidth. But easy conformability on a curved surface is its best feature. The natural resonance modes under a triangular patch and the modal nomenclature are a bit different from those occurring under a circular or rectangular geometry. They are never symmetric with respect to the H-plane while resonating with the dominant $TM_{1,1,-1}$ mode [32]. A recent study [33] has pointed out a remarkable observation that no higher order mode is responsible in generating the XP radiation from a triangular patch. Unlike a rectangular or circular geometry, the primary resonant mode itself acts as the source of the cross-polar fields [33]. Simple symmetric and linear DGS parallel to the E-plane was tried in [34] demonstrating about 4–5 dB suppression in the H-plane XP level. But that design is too sensitive to achieve the required fabrication tolerances. A better design is possible just by bending one end of the DGS in the form of "L." The scheme is shown in Figure 4.24. Its radiation performance is shown in Figure 4.25 revealing 5–6 dB suppression in peak XP level.

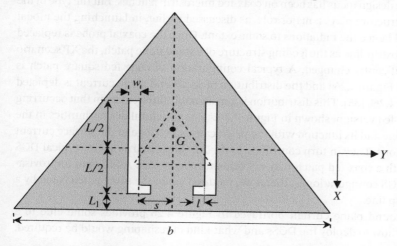

Figure 4.24 An equilateral triangular patch integrated with "L"-shaped DGS (bottom view).

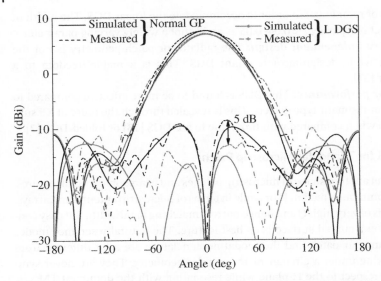

Figure 4.25 H-plane radiation patterns of a triangular patch with and without L-shaped DGS (as in Figure 4.24), $f = 6.2$ GHz. substrate thickness = 1.575 mm, $\varepsilon_r = 2.33$, sides length = 19 mm, probe location 9.1 mm from the vertex, $b = 100$ mm, $L = 30$ mm, $w = 3.5$ mm, $s = 8.5$ mm, $l = 1.5$ mm.

4.5 Suppression of Cross-Polarized Fields by DGS Integration Technique: Microstrip-Fed Patches

So far, the design focus has been on coax-fed microstrip patches. But the type of the feeding structure plays a major role, as discussed earlier, in launching the modal fields and hence the radiations to some extent. Once the coaxial probe is replaced by a microstrip-line as the feeding structure of a standalone patch, the XP scenario gets significantly changed. A typical configuration of edge fed square patch is shown in Figure 4.26a and the distribution of its ground plane current is depicted in Figure 4.26b [35]. This distribution is considerably different from that occurring in a coax-fed version, shown in Figure 4.26c. The structural discontinuities in the feeding line and its junction with the patch causes a change in the surface current distributions. They in turn contribute to the XP values and as such a typical DGS tested with a coax-fed patch may not remain valid in toto. Additional improvisation in DGS configuration is, therefore, required when the same patch is fed by a microstrip line.

The ground plane current portrayed in Figure 4.26 provides some clue to a designer how to deploy the DGSs and what kind of reshaping would be required.

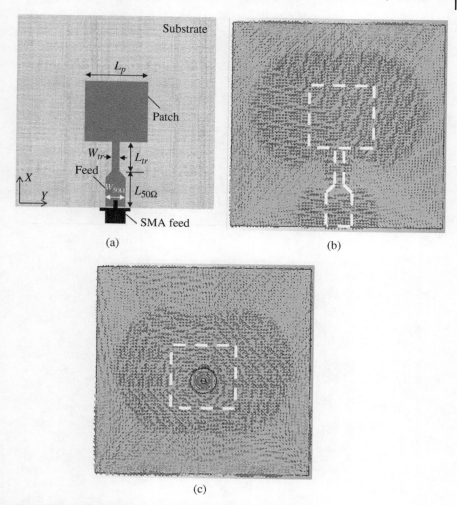

Figure 4.26 A square patch with different feed configurations: (a) layout with a microstrip line feed; (b) simulated ground plane current due to microstrip-fed patch; (c) simulated ground plane current when the same patch is excited by a coaxial feed. Source: Pasha et al. [35] © IEEE.

Figure 4.27a gives a clear picture. DGS pair D_1D_1 takes care of the higher order TM_{02} mode and an additional pair D_2D_2 has been conceived to take care of the radiations from the discontinuity in the feed itself [35]. Indeed, such discontinuity in a feed line is very common and is frequently required for the sake of impedance matching. But it produces a special kind of radiation which is similar to that

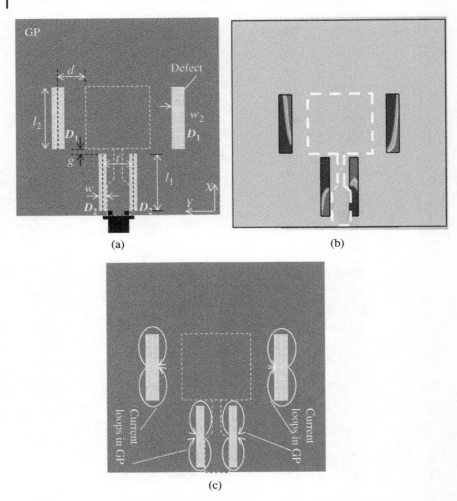

Figure 4.27 DGS integrated microstrip-fed patch: (a) layout from the bottom side; (b) induced magnetic fields in the DGS slots; (c) schematic diagram indicating current loops around the DGSs. Source: Pasha et al. [35] © IEEE.

caused by a magnetic dipole [36]. Therefore, a counter magnetic dipole might help in nullifying this unwanted radiation from the line discontinuity. Actually, D_2D_2 DGS segments serve that purpose.

The induced magnetic fields inside the DGS slots are shown in Figure 4.27b. Figure 4.27c depicts a simplified sketch of the ground plane current around those slots which resembles magnetic dipoles. The DGS slots D_1D_1 also interact with the spurious radiations in an identical manner. The measured radiation patterns of this antenna are shown in Figure 4.28. No significant difference due to the integration of the DGS is revealed in the co-polarized patterns. But, the XP isolation over the H-plane endorses about 15 dB improvement.

Figure 4.28 Radiation patterns at resonance for the antennas in Figure 4.27 with and without DGS: (a) E-plane; (b) H-plane. Source: Pasha et al. [35] © IEEE.

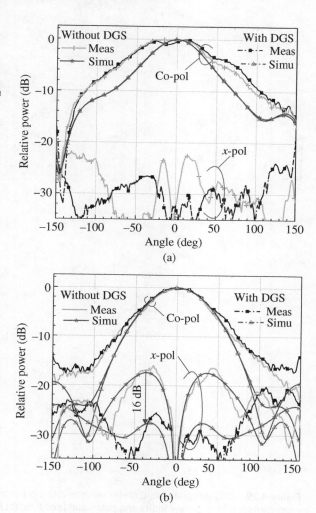

A variant of folded DGS explored in a probe-fed patch [7] has been successfully employed for a millimeter wave sensor design at 28 GHz [37]. The antenna layout [37] is shown in Figure 4.29 which depicts a pair of additional horizontal sections in addition to "[" and "]". This antenna operating over 5.35% matching bandwidth reveals more than 14 dB increase in the XP isolation [37].

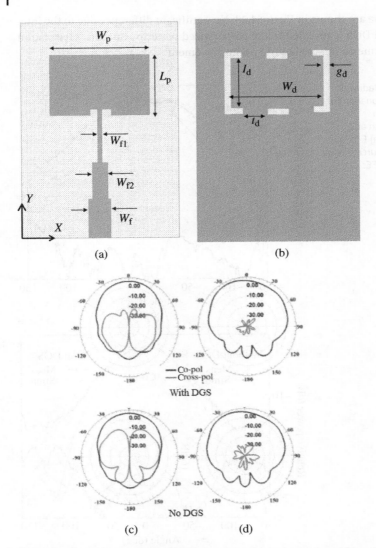

Figure 4.29 DGS integrated millimeter wave microstrip patch and its radiation characteristics: (a) top view indicating patch and feed line; (b) DGS viewed from ground plane side; (c) H-plane radiation patterns; (d) E-plane radiation patterns. Source: Rezaei et al. [37].

4.6 Recent Works and New Trends

4.6.1 New DGS Geometries

One of the trends is to use different variations in DGS shapes with a view to reduce the size and increase in XP suppression. A few examples are shown in Figure 4.30 which embodies relatively wide slot [38] and dumbbell-shaped [39] defects. Both shapes were investigated for coax-fed rectangular microstrips. It is important to note that, the working principle for these designs remains same as the earlier ones [7, 31], i.e. to weaken TM_{02} mode. These configurations provide moderate suppressions in XP fields over the H-plane ranging from 10 to 12 dB.

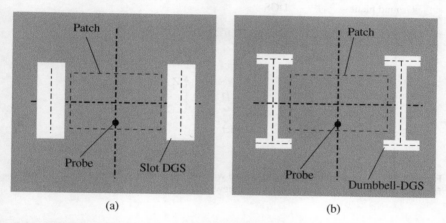

(a)   (b)

Figure 4.30 Coax-fed rectangular patch integrated with (a) slot-DGS [38] and (b) dumbbell shaped DGS. Source: Adapted from Ghosh et al. [39].

4.6.2 New Design Concept of Substrate Field Symmetry

A different approach in conceiving the DGS integration is equally interesting. One striking relation between the substrate E-fields and XP radiation was first reported in [31]. That investigation was executed with both circular and rectangular patches. That work established a strong relation between the asymmetry in substrate fields and high cross-polarized radiations as discussed in Section 4.4.4 [7, 28]. Obtaining "symmetry" in the field distribution is a major indicator behind realizing improved polarization purity. Thus, the DGS-integration may be considered as a means of redistributing the field and gaining its symmetry. This approach has been successfully implemented in several designs where the DGSs have been deployed even far apart from the patch edge. They hardly can find any scope of direct interaction with the higher modes. Figure 4.31 shows one such DGS integrated design [40] where non-proximal symmetrical DGS around the

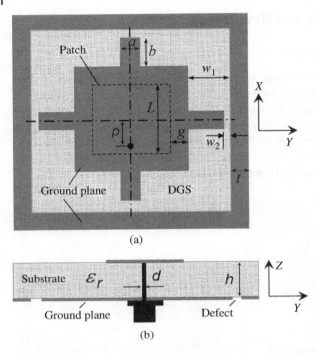

(a)

(b)

Figure 4.31 A rectangular patch integrated with non-proximal symmetric DGS: (a) view from the ground plane; (b) cross-sectional view. Source: Kumar et al. [40] © IEEE.

patch contour has been employed. The resulting changes in substrate fields along with the radiation characteristics are also documented in Figure 4.32.

A similar design concept was implemented by an asymmetrically shaped DGS like "*L*" as shown in Figure 4.33 [41, 42]. The idea is to address the field asymmetry around the patch by a DGS with asymmetric shape. The performance of asymmetric DGS compared to the other symmetrical configurations discussed so far is shown in Figure 4.34. The asymmetric "*L*" DGS provides maximum XP suppression among all. The linear resonant DGS [31] may be closely comparable in terms of performance; however, its deployment area is about 2.5 times larger compared to an *L*-DGS. The same observation was made with the asymmetric-arc DGS for circular patch [43]. This mechanism reduces the "defect" area significantly and leaves more room for other circuits, connectors, or components. A comparative study for those different types of DGSs integrated with rectangular patches is presented in Table 4.1. This helps one in understanding the variations in performance as functions DGS geometry as well as the required area of deployments.

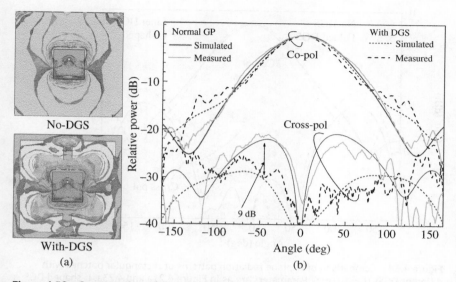

Figure 4.32 Studies of the effect of DGS in the design of Figure 4.31: (a) simulated substrate fields; (b) H-plane radiation patterns. Source: Kumar et al. [40] © IEEE.

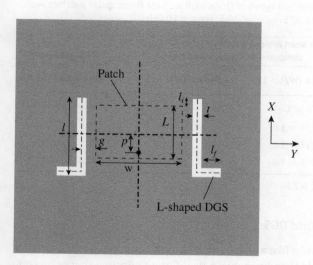

Figure 4.33 L-shaped DGS integrated rectangular patch. Source: Kumar and Guha [42].

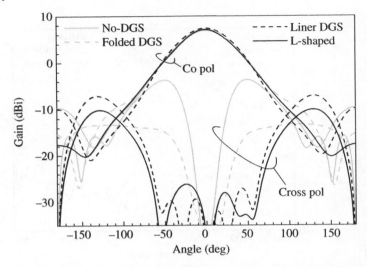

Figure 4.34 Comparison of H-plane radiation patterns of rectangular patches with different DGSs ($f = 10$ GHz). Parameters are as in Figure 4.21a and 4.23a. L-shaped DGS parameters: $l = 9.35$ mm, $l_f = 3.1$ mm, $l_i = 0$ mm, $g = 0$ mm, $t = 1.5$ mm. Source: Kumar and Guha [41] © IEEE.

Table 4.1 Comparison of different types of DGSs explored for rectangular patches with different W/L Substrate: $h = 1.575$ MM; $\varepsilon_r = 2.33$; Freq. ≈ 10 GHz.

DGS type	Angular span around boresight with XP < -25 dB compared to co-polarized peak		DGS Area (mm^2)	
	Rectangle ($W/L > 1$)	Square ($W/L = 1$)	Rectangle	Square
Folded [7]	86° ($W/L = 1.6$)	123°	48.9	53.4
Slot [38]	105° ($W/L = 1.5$)	—	760	—
Linear [31]	150° ($W/L = 1.6$)	161°	90	90
Asymmetric [42]	162° ($W/L = 1.6$)	164°	37.4	41.8

Source: Kumar and Guha [42] IEEE.

4.6.3 Reconfigurable Grid DGS

Conventionally, a DGS looks like a shaped slot. Another concept is to construct a DGS from a grid of slots [44] and to achieve its different effective shapes by controlling "open" or "short" using appropriate hardwire. A typical example is shown in Figure 4.35a where a pair of grid DGSs is visible from the back side [44]. It may be considered as a series of square metallic islands systematically arranged within

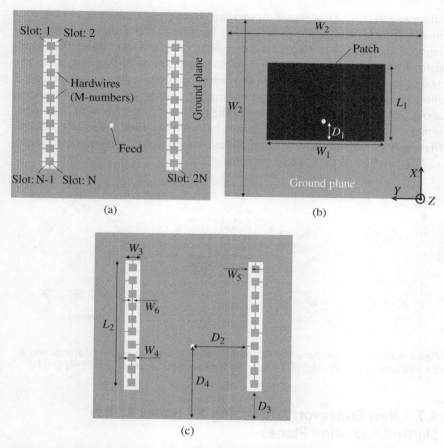

Figure 4.35 (a) General configuration of a gridded DGS (GDGS) (b) top view of a rectangular patch (c) configuration of a GDGS applied to the patch for reduction of the XP radiations. Source: Zhang et al. [44] © IEEE.

a rectangular slot where each island is connected to the main ground or to the adjacent ones by thin metallic interconnects. Thus, it eventually creates a series of narrow slots separated by those interconnects. If any of them is removed, two adjacent slots get merged and a different shape is generated. This is the idea of configurability in binary mode with value "0" for short using interconnect and value "1" for open through removing the interconnect.

In the configuration of Figure 4.35a, the pair of grids contains 2×22 (=M) interconnects and may result in 2^M combinations of DGS shapes. An artificial intelligence (AI) based optimization routine may be developed to find out the best possible configuration to realize the target XP level. This grid DGS is suitable for either suppressing XP radiation or exciting circular polarization (CP) [44]. As a matter of fact, same logic can be used to increase the XP component in realizing a CP antenna. More details are discussed in Chapter 8. An S-band rectangular patch integrated with a pair of grid DGSs [44] is shown in Figure 4.35b,c. The grid geometry of the DGS has been kept mirror symmetric about y-axis since it is supposed to interact with the XP generating TM_{02} mode which is also symmetric about this y-axis. Figure 4.36 compares the radiation properties with its conventional counterpart. No significant change in the co-polarized radiations confirms that the dominant mode remains unaffected. However, a suppression of the order 15 dB is revealed in the H-plane XP fields.

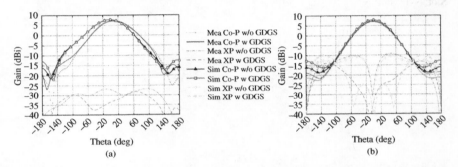

Figure 4.36 Radiation characteristics of the antenna shown in Figure 4.35, obtained at the mid band ($f = 3.65$ GHz): (a) E-plane; (b) H-plane. Source: Zhang et al. [44] © IEEE.

4.7 New Endeavor: Addressing XP Issues Across Skewed Radiation Planes

So far, all investigations were primarily focused to the orthogonal plane XP issues and the remedial by DGS integration method. This orthogonal plane actually indicates its H-plane which is an obvious part of the principal radiation planes. Commonly the antenna engineers characterize an antenna radiation across the principal planes which mean both E- and H-planes. Sometimes during the measurements, they are referred to as vertical cut and horizontal cut. But the radiation indeed is a three-dimensional (3D) phenomenon. So, there are multiple planes in between E- and H-planes which are called skewed radiation planes. A set of diagonal planes belongs to this category and according to the theoretical analysis [45], the cross-polar radiation should attain the maximum value over these diagonal planes (azimuth angle $\varphi = 45°$ from the nearest E-plane or H-plane). But in the standard mode of characterization, we frequently ignore these planes. In reality, a microstrip antenna behaves a bit differently as examined in Figure 4.37 [7]. It shows the variation of XP level of a coax-fed rectangular patch as a function of the radiation plane [7]. Three different aspect ratios (W/L) have been considered and interestingly, in all cases, the peak XP occurs near 70° instead of the theoretical prediction of having it around 45°. An asymmetry in the feed location with respect to the patch center is the most probable reason behind this observation. Now the question is: *could DGS work and take care of suppressing the XP fields over the skewed radiation planes?*

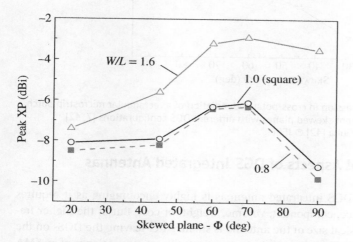

Figure 4.37 Cross-polarized isolation of probe fed rectangular microstrip patches for different skewed planes. Source: Kumar and Guha [7] With permission of Institution of Engineering and Technology.

This question has been addressed in [7, 42] and a set of representative results is shown through Figure 4.38. This is for a rectangular patch with $W/L = 1.6$ and different types of DGSs have been used as are clearly labeled inside the figure. The order of suppression has been plotted against the radiation planes. It is insignificant near the E-plane ($\Phi \approx 0°$) where XP level itself is very weak and shows its maximum effect near H-plane ($\Phi = 90°$) where the natural XP level is moderate. But around $\Phi \approx 70°$ where XP is a maximum, the order of suppression by DGS is about 10 dB. The same occurring in $\Phi \approx 90°$ is about 10–15 dB.

The handling the XP radiations over the diagonal or skewed planes is a bit tricky. That needs more detailed understanding and information about the multi-parametric sources of XP radiations, which are not commonly known to the antenna community. The advanced information and the related DGS techniques have been exclusively addressed in Chapter 5.

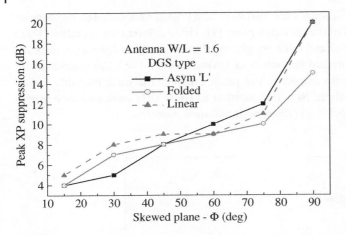

Figure 4.38 Suppression in cross-polarized radiation of a rectangular microstrip patch (*W*/*L* = 1.6) at different skewed planes with different DGS configurations [7, 42]. Source: Kumar and Guha [42] © IEEE.

4.8 Practical Aspects of DGS-Integrated Antennas

The potential of DGS integrated antennas is highly encouraging as it requires no additional space, component, volume, weight, or cost. But at the higher frequencies the physical size of the antenna is small. After leaving the DGSs on the ground, it may be difficult to get adequate room for the standard flange of an SMA

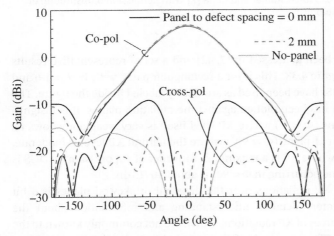

Figure 4.39 H-plane radiation patterns of circular patches with and without arc-DGS when placed on a flat metallic panel with variable intermediate spacing. Parameters as in Figure 4.14.

connector to be fitted. Hence some modification in connector body is called for. This seems to be a major limitation for a probe fed design and the limiting range of usable frequency. However, the non-proximal DGS [40] or asymmetric-DGS [42] leaves adequate space for mounting the SMA connector and resolves such issues.

Another issue may arise when one needs to mount the antenna on a metallic support. In that case, the DGS would get shorted by the additional metallic surface touching the ground plane. To avoid this crisis, a secondary ground plane beneath the primary one maintaining a suitable mutual separation could be an acceptable solution [46]. A study with a circular patch bearing arc-shaped DGS has been demonstrated in Figure 4.39. When the antenna is placed directly on a metal plate, the DGS does not work as reveled by its XP level. As soon as it is separated is by 2 mm $(0.06 \lambda_0)$ from the primary ground plane, the DGS becomes effective. The XP performance is quite acceptable and comparable with that obtained without any secondary metal behind it.

References

1 R. Garg, P. Bhartia, I. Bahl, and A. Ittipiboon, "*Microstrip Antenna Design Handbook,*" Boston, *Artech House*, 2001.

2 C. A. Balanis, "*Antenna Theory Analysis and Design,*" John Wiley & Sons Inc, New Jersey, 2005.

3 J. Huang, "The finite ground plane effect on the microstrip antenna radiation patterns," *IEEE Transactions on Antennas and Propagation*, vol. 31, no. 4, pp. 649–635, 1983.

4 T. Huynh, K. F. Lee, and R. Q. Lee, "Cross polarization of rectangular patch antennas," *Electronic Letters*, vol. 24, no. 8, pp. 463–464, 1988.

5 D. Guha, "Microstrip antenna cross-polarized radiations, Kai Fong Lee, and recent insightful observations," *IEEE Asia Pacific Conference on Microwave*, pp. 375–376, 8–11 December 2020, Hong Kong.

6 A. C. Ludwig, "Definition of cross polarization," *IEEE Transactions on Antennas and Propagation*, vol. 21, no. 1, pp. 116–119, 1973.

7 C. Kumar and D. Guha, "Defected ground structure (DGS)-integrated rectangular microstrip patch for improved polarization purity with wide impedance bandwidth," *IET Microwaves, Antennas & Propagation*, vol. 8, no. 8, pp. 589–596, 2014.

8 M. L. Oberhart, Y. T. Lo, and R. Q. H. Lee, "New simple feed network for an array module of four microstrip elements," *Electronics Letters*, vol. 23, no. 9, pp. 436–437, 1987.

9 S. Biswas, D. Guha, and C. Kumar, "Control of higher harmonics and their radiations in microstrip antennas using compact defected ground structures," *IEEE Transactions on Antennas and Propagation*, vol. 61, no. 6, pp. 3349–3353, 2013.

10 K. F. Lee, K. M. Luk, and P. Y. Tam, "Cross-polarization characteristics of circular patch antennas," *Electronics Letters*, vol. 28, no. 6, pp. 587–589, 1992.

11 D. Guha, M. Biswas, and Y. M. M. Antar, "Microstrip patch antenna with defected ground structure for cross polarization suppression," *IEEE Antennas and Wireless Propagation Letters*, vol. 4, pp. 455–458, 2005.

12 C. Kumar and D. Guha, "A new look into the cross-polarized radiation form a circular microstrip antenna and suppression using dot-shaped DGS," *IEEE International Symposium on Antennas & Propagation*, pp. 1–4, 11–17 July 2010, Toronto, Canada.

13 C. Kumar and D. Guha, "Nature of cross-polarized radiations from probe-fed circular microstrip antennas and their suppression using different geometries of defected ground structure (DGS)," *IEEE Transactions on Antennas and Propagation*, vol. 60, no. 1, pp. 92–101, 2012.

14 Q-R. Zheng, Y-Q. Fu, and N-C. Yuan, "A novel compact spiral electromagnetic band-gap (EBG) structure," *IEEE Transactions on Antennas and Propagation*, vol. 56, no. 6, pp. 1656–1660, 2008.

15 A. Petosa, A. Ittipiboon, and N. Gagnon, "Suppression of unwanted probe radiation in wideband probe-fed microstrip patches," *Electronics Letters*, vol. 35, no. 5, pp. 355–357, 1999.

16 V. Schejbal and V. Kovarik, "A method of cross-polarization reduction," *IEEE Antennas and Propagation Magazine*, vol. 48, no. 5, pp. 108–111, 2006.

17 Z. N. Chen and M. Y. W. Chia, "Broad-band suspended probe-fed plate antenna with low cross-polarization level," *IEEE Transactions on Antennas and Propagation*, vol. 51, no. 2, pp. 345–346, 2003.

18 W. H. Hsu and K. L. Wong, "Broad-band probe-fed patch antenna with a U-shaped ground plane for cross-polarization reduction," *IEEE Transactions on Antennas and Propagation*, vol. 50, no. 3, pp. 352–355, 2002.

19 K. L. Wong, C. L. Tang, and J. Y. Chiou, "Broad-band probe-fed patch antenna with a W-shaped ground plane," *IEEE Transactions on Antennas and Propagation*, vol. 50, no. 6, pp. 827–831, 2002.

20 X. Zhang and L. Zhu, "Patch antennas with loading of a pair of shorting pins toward flexible impedance matching and low cross polarization," *IEEE Transactions on Antennas and Propagation*, vol. 64, no. 4, pp. 1226–1233, 2016.

21 C. Kumar and D. Guha, "Higher mode discrimination in a rectangular patch: new insight leading to improved design with consistently low cross-polar radiations," *IEEE Transactions on Antennas and Propagation*, vol. 69, no. 2, pp. 708–714, 2021.

22 D. Dutta, D. Guha, and C. Kumar, "Mitigating unwanted mode in a microstrip patch by a simpler technique to reduce cross-polarized fields over the orthogonal plane," *IEEE Antennas Wireless Propagation Letters*, vol. 20, no. 5, pp. 678–682, 2021.

23 D. Dutta, D. Guha, and C. Kumar, "Microstrip patch with grounded spikes: a new technique to discriminate orthogonal mode for reducing cross-polarized radiations," *IEEE Transactions on Antennas and Propagation*, vol. 70, no. 3, pp. 2295–2300, 2022.

24 High Frequency Structure Simulator (HFSS), Version 11.1, Ansoft, Pittsburgh, PA, USA, 2008.

25 C. Kumar and D. Guha, "New defected ground structures (DGSs) to reduce cross-polarized radiation of circular microstrip antennas," *IEEE Applied Electromagnetic Conference*, pp. 1–4, 14–16 December 2009, Kolkata, India.

26 D. Guha, C. Kumar, and S. Pal, "Improved cross-polarization characteristics of circular microstrip antenna employing arc-shaped defected ground structure (DGS)," *IEEE Antennas Wireless Propagation Letters*, vol. 8, pp. 1367–1369, 2009.

27 C. Kumar and D. Guha, "Linearly polarized elliptical microstrip antenna with improved polarization purity and bandwidth characteristics," *Microwave and Optical Technology Letters,* vol. 54, no. 10, pp. 2309–2314, 2012.

28 C. Kumar and D. Guha, "Modulation of substrate fields: key to realize universal DGS configuration for suppressing cross-polarized radiations from a microstrip patch having any geometry," *IEEE Antennas and Propagation Society International Symposium Digest*, pp. 1–2, 8–13 July 2012, Chicago, IL, USA.

29 CST Studio Suite® 2019, Dassault Systemes India Pvt. Ltd, New Delhi, India, 2019.

30 "Feldberechnung für körper mit beliebiger oberfläche (FEKO)," Altair Engineering, Troy, Michigan, United States, 2020.

31 C. Kumar and D. Guha, "Reduction in cross-polarized radiation of microstrip patches using geometry independent resonant-type defected ground structure (DGS)," *IEEE Transactions on Antennas and Propagation*, vol. 63, no. 06, pp. 2767–2772, 2015.

32 J. S. Dahele and K. F. Lee, "On the resonant frequencies of the triangular patch antenna," *IEEE Transactions on Antennas and Propagation*, vol. 35, no. 1, pp. 100–101, 1987.

33 C. Sarkar, D. Guha, and C. Kumar, "Source of cross-polar fields in a triangular patch: insight and experimental proof," *IEEE Antenna Wireless Propagation Letters,* vol. 20, no. 12, pp. 2437–2441, 2021.

34 P. Maity, C. Kumar, and D. Guha, "Triangular patch with linear DGS for improved polarization purity," *Proceeding of IEEE Applied Electromagnetic Conference-AEMC*, December 2013, Bhubaneswar.

35 M. I. Pasha, C. Kumar, and D. Guha, "Simultaneous compensation of microstrip feed and patch by defected ground structure for reduced cross-polarized radiation," *IEEE Transaction on Antennas and Propagation,* vol. 66, no. 12, pp. 7348–7352, 2018.

36 L. Lewin, "Radiation from discontinuities in strip-line," *Proceedings of the IEE-Part C: Monograph,* vol. 107, no. 12, pp. 163–170, 1960.

37 M. Rezaei, H. Zamani, M. Fakharzadeh, and M. Memarian, "Quality improvement of millimeter-wave imaging systems using optimized dual polarized arrays," *IEEE Transactions on Antennas and Propagation,* vol. 69, no. 10, pp. 6848–6856, 2021.

38 A. Ghosh, D. Ghosh, S. Chattopadhyay, and L. Singh, "Rectangular microstrip antenna on slot type defected ground for reduced cross polarized radiation," *IEEE Antennas Wireless Propagation Letters,* vol. 14, pp. 321–324, 2015.

39 A Ghosh, S. Chakraborty, S. Chattopadhyay, A. Nandi, and B. Basu, "Rectangular microstrip antenna with dumbbell shaped defected ground structure for improved cross polarised radiation in wide elevation angle and its theoretical analysis," *IET Microwaves, Antennas & Propagation,* vol. 10, no. 1, pp. 68–78, 2016.

40 C. Kumar, M. I. Pasha, and D. Guha, "Microstrip patch with non-proximal symmetric defected ground structure (DGS) for improved cross-polarization properties over principal radiation planes," *IEEE Antennas Wireless Propagation Letters,* vol. 14, pp. 1412–1414, 2015.

41 C. Kumar and D. Guha, "L-shaped defected ground structure: small in size but significant in suppressing cross-polarized fields," *5th IEEE Applied Electromagnetics Conference (AEMC),* pp. 1–2, December 2015, Guwahati, India.

42 C. Kumar and D. Guha, "Asymmetric geometry of defected ground structure for rectangular microstrip: a new approach to reduce its cross-polarized fields," *IEEE Transactions on Antennas and Propagation,* vol. 64, no. 06, pp. 2503–2506, 2016.

43 C. Kumar and D. Guha, "Asymmetric and compact DGS configuration for circular patch with improved radiations," *IEEE Antennas Wireless Propagation Letters,* vol. 19, no. 2, pp. 355–357, 2020.

44 Y. Zhang, Z. Han, S. Shen, C. Y. Chiu, and R. Murch, "Polarization enhancement of microstrip antennas by asymmetric and symmetric grid defected ground structures," *IEEE Open Journal of Antennas and Propagation,* vol. 1, pp. 215–223, 2020.

45 R. C. Hansen, "Cross polarization of microstrip patch antennas," *IEEE Transactions on Antennas and Propagation,* vol. 35, no. 6, pp. 731–732, 1987.

46 D. Guha, S. Biswas, M. Biswas, J. Y. Siddiqui, and Y. M. M. Antar, "Concentric ring shaped defected ground structures for microstrip circuits and antennas," *IEEE Antennas Wireless Propagation Letters,,* vol. 5, pp. 402–405, 2006.

5

Multi Parametric Cross-Polar Sources in Microstrip Patches and DGS-Based Solution to All Radiation Planes

5.1 Background and Introduction

The microstrip antenna was conceptually realized in 1953 [1] and tested as a practical radiator in the early 1970s [2, 3]. Till the early 1980s, the main focus was studying the feed, bandwidth, and radiation properties without any attention to the issue of cross-polarized (XP) fields [4]. Hansen [5] first analyzed XP radiations from square and circular patches on the basis of far fields "produced by the dominant cavity mode" [6]. No definite idea about the source of those XP radiations was available that time. An investigation with lateral strip fed rectangular microstrip array first accounted for the orthogonal modes and considered them as XP source [7]. More convincing studies by Lee and coworkers [8, 9] established that observation through analysis and experiments. Ludwig's definitions on XP fields [10] were the basis of expressing both co- and cross-polarized fields in [9] but there was no comment on the actual number of the modes taking part in XP generation. Later on, it is found the first higher order mode becomes responsible and in some cases like circular or square patches, orthogonally oriented dominant mode or degenerate mode could be yet another source [11].

Conventionally, the XP values are discussed across the principal radiation planes and predominantly in the H-plane. This is based on a notion that orthogonally polarized XP fields would be more aligned to the orthogonal plane or H-plane. But the reality is different as was first indicated in [9]. The XP level actually attains the maximum value over the diagonal planes (D-planes) although it is frequently ignored in the typical investigations and technical articles.

The insightful investigations in [12, 13] had successfully discriminated the so called orthogonal higher mode in a rectangular patch. That benefitted in enormously in reducing the XP level but only over the H-plane. Chapter 4 indeed has focused on that aspect of DGS-based designs. But more significantly, such higher mode discrimination impacts no change in the diagonal plane XP scenario. It is

Defected Ground Structure (DGS) Based Antennas: Design Physics, Engineering, and Applications,
First Edition. Debatosh Guha, Chandrakanta Kumar, and Sujoy Biswas.
© 2023 The Institute of Electrical and Electronics Engineers, Inc. Published 2023 by John Wiley & Sons, Inc.

therefore an established fact that the orthogonal higher or degenerate mode is not the sole source of XP radiations.

More advanced investigations in search of the XP sources have resulted in several new information [14, 15] which are useful from the practical point of view. They enable one in mitigating the odd situations and achieving high polarization purity uniformly over the entire radiation planes for advanced design. A few DGS-based approaches [15–17] have been discussed.

5.2 Mathematical Explanations of Cross-Polarized Fields

A study [14] has revisited the sources in terms of near E_x and E_y and produced a mathematical model using Ludwig's third definition [10]. That tried to correlate both H- and D-plane XP fields C^H and C^D respectively with the near E field components. With reference to a typical microstrip radiator shown in Figure 5.1 they can be expressed as [17]

$$C^H = \frac{j\beta e^{-j\beta r}}{4\pi r} \iint 2E_y e^{j\beta \mathbf{r}' \cdot \hat{\mathbf{r}}} ds' \tag{5.1}$$

$$C^D = \frac{j\beta e^{-j\beta r}}{4\pi r} \iint \{E_y[1 + \cos\theta] - E_x[1 - \cos\theta]\} e^{j\beta \mathbf{r}' \cdot \hat{\mathbf{r}}} ds' \tag{5.2}$$

where, $\mathbf{r}'$ is the source point from the origin and $\hat{\mathbf{r}} = \frac{\mathbf{r}}{|\mathbf{r}|}$ is the unit vector towards the direction of observation. This pair of equations indicates that H-plane XP is a sole function of E_y; however, both E_x and E_y contribute to D-planes.

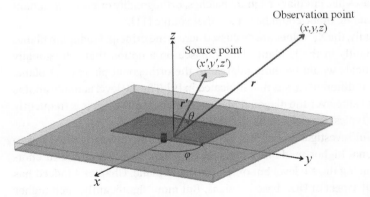

Figure 5.1 A probe-fed radiating microstrip patch antenna.

5.2.1 Sources of E_x and E_y Components

The question is where do E_x and E_y come from? Figure 5.2 helps in understanding the same since it schematically portrays the modal fields of two main players, i.e. TM_{10}, the radiating mode and TM_{02}, the most probable XP generating mode in a rectangular microstrip element. An isolated patch behaves as a partially open cavity with perfect electric walls at the top and at the bottom and magnetic walls surrounding the edges. Therefore, TM_{10} mode reveals zero field at the patch center at $x = 0$ and maximum at the patch edges with $x = \pm L/2$. Following the same boundary conditions, TM_{02} mode peaks at the patch edges along with a maximum at the patch center. Their contributions to radiation are schematically portrayed in Figure 5.3 which is self-explanatory. Once the patch starts resonating, its vertical E_z fields fringe out near the open boundary and remains no longer purely vertical. The inclined field vectors comprise both horizontal and vertical components. If one looks from top, the effective fields revealed to that viewer contain only the horizontally polarized E_x fields. The associated magnetic field would be simply H_y and thus they cause radiation along z axis as per the Poynting vector P_z. The same narrative for TM_{02} mode with respect to

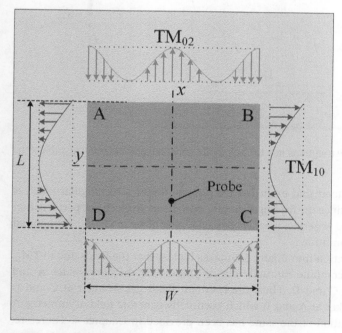

Figure 5.2 Modal electric field distributions for TM_{10} and TM_{02} modes beneath a rectangular microstrip patch.

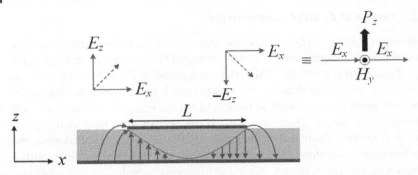

Figure 5.3 E-field distributions due to TM_{10} under a resonant microstrip patch.

Figure 5.4 ideally should result in radiation null in the boresight (along z axis). But $E_{y,TM02}$ across the nonradiating edges radiate obliquely over the orthogonal or H-plane.

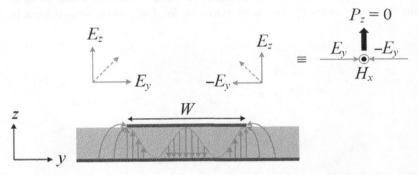

Figure 5.4 E-field distributions due to TM_{02} mode under a rectangular microstrip patch.

It is well understood that exclusive TM_{10} mode under a rectangular patch is next to impossible since the feeding structure introduces an electrical perturbation inside the cavity and generates weakly coupled higher order mode(s), as supported by the boundary condition.

A careful observation into Figure 5.2 makes it clear that the fields due to TM_{10} and TM_{02} modes combine mutually out of phase near the coordinates A and B, but in phase at C and D. Thus, the resultant fields get stronger at C and D and relatively weaker at A and B which eventually generate field asymmetry. A detailed demonstration of the same has been provided in Chapter 4. Therefore, $E_{y,TM02}$ components do not fully cancel out and a weak residual E_y takes part in XP radiations in the boresight which is commonly observed in a practical radiator.

5.2.2 How to Combat E_y Components

A traditional microstrip patch possesses high XP radiation, although its intensity varies with the element geometry and the feed configuration. Based on the existing information, the antenna researchers have tried to mitigate the issues in a variety of ways that include balanced feed [18], ground plane shaping [19], shaping of probe [20], and dual-layer substrate [21]. Another simple and low-profile technique has been known since 2005 [11] which employ defects on the ground [22–24].

An investigation has exclusively targeted TM_{02} [12] and has encountered it strategic shorting pins across the non-radiating edges of a rectangular patch. The geometry is shown in Figure 5.5. The shorting pins are supposed to block $E_{y,TM02}$ and its effect is manifested through the H-plane radiation patterns as shown in Figure 5.6a. The peak XP level decreases by more than 15 dB. But a significant observation is ushered by Figure 5.6b. The discrimination of $E_{y,TM02}$ hardly causes any change in the D-plane XP values. This fact is endorsed by another work [13] which uses a pair of grounded spikes as shown in Figure 5.7. Here also, the purpose of the spikes is to weaken TM_{02} mode and that is reflected in Figure 5.8a. The order of XP suppression in the H-plane is more than 20 dB although the situation over the D-plane (Figure 5.8b) is extremely disappointing. It is important to note that both designs [12, 13] significantly weaken TM_{02} mode (i.e. E_y fields) but maintain the primary radiation and co-polar gain (i.e. E_x fields) unaffected.

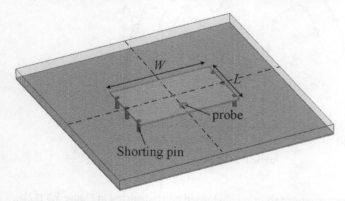

Figure 5.5 A coax-fed rectangular microstrip patch with shorting pins along the non-radiating edges. Source: Adapted from Kumar and Guha [12].

According to (5.1) and (5.2), both H- and D-plane XP should be reduced when E_x remains same and E_y gets weakened. But this did not happen in [12, 13]. This, in turn, raises a question: *does there exist any additional sources other than the known parameters such as E_x and E_y?*

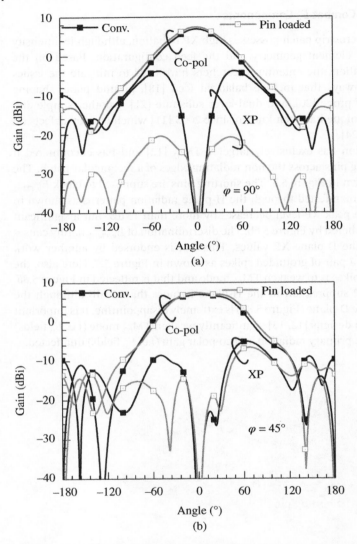

Figure 5.6 Radiation patterns of shorting pin loaded patch depicted in Figure 5.5 [12]: (a) H-plane; (b) D-plane. Source: Adapted from Kumar and Guha [12].

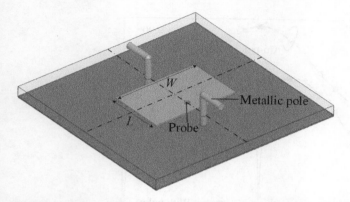

Figure 5.7 A rectangular patch with a pair of grounded metallic spikes near its radiating edges. Source: Dutta et al. [13] © IEEE.

5.3 Detailed Investigations in to the XP Sources

It is difficult to attain a complete solution of XP radiations over full azimuth without having a comprehensive set of information about the source fields. The investigations in [15] are exhaustive based on three representative patch geometries as shown in Figure 5.9. The target fields such as E_x, E_y, H_x, and H_y have been considered as a function of φ on the upper surface of the substrate at a radial distance S. Its value is arbitrarily chosen as $S = 1.25L$ for rectangular and square patches, and $S = 2R$ for circular geometry. Each radiating element has been fed by four different feeding techniques, e.g. coaxial probe (designated as feed#1), microstrip inset feed (feed#2), microstrip edge feed (feed#3), and aperture coupled feed (feed#4), and all necessary data have been generated using commercial simulation tool [25]. This huge volume of surface field data is processed and analyzed through Figures 5.10–5.14.

The field information has been collected for φ values at 5° interval, but for the sake of the study they have been organized across three representative planes with $\varphi = 0°$, 45°, and 90°. Figures 5.10–5.14 also embody corresponding XP levels which are simply marked by hollow bars.

5.3.1 Rectangular Patch

Figure 5.10 is dedicated to rectangular patch geometry and it examines any correlations between the surface fields and corresponding XP generation for each feed configuration across three specific radiation planes with $\varphi = 0°$ (E-plane), $\varphi = 45°$ (D-plane), and $\varphi = 90°$ (H-plane).

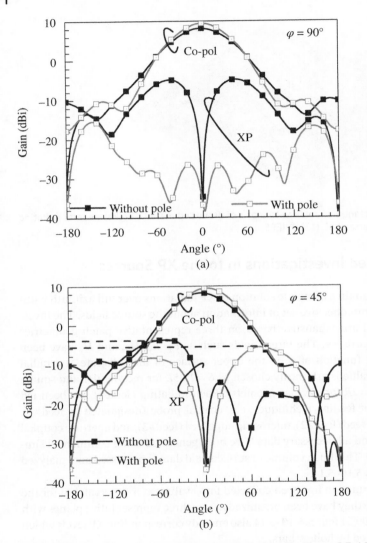

Figure 5.8 Radiation patterns of the antenna depicted in Figure 5.7: (a) H-plane; (b) D-plane [13]. Source: Dutta et al. [13] © IEEE.

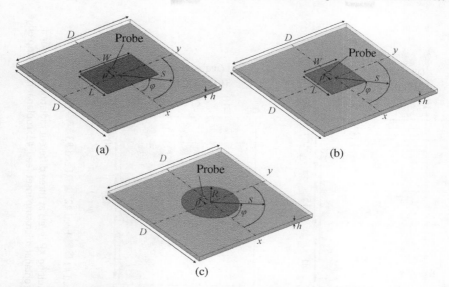

Figure 5.9 Probe-fed conventional patches: (a) rectangular; (b) square; (c) circular. Parameters: $h = 1.575$, $\varepsilon_r = 2.33$, $D = 50$; rectangular: $L = 15$, $W = 22.5$, $\rho = 4$, $S = 1.25 L$; square: $L = W = 15$, $\rho = 2.6$, $S = 1.25 L$; circular: $R = 9.5$, $\rho = 2.7$, $S = 2R$, all dimensions in millimeters. S is the radius of the arc – an imaginary line of observation [15]. Source: Rafidul et al. [15] © IEEE.

The observations for E_x in Figure 5.10 is as follows: The highest value of $E_{x(\varphi \sim 45°)}$ for feed #4 corresponds to the lowest $XP_{(\varphi \sim 45°)}$; but the lowest $E_{x(\varphi \sim 45°)}$ for feed #2 does not correspond to the highest $XP_{(\varphi \sim 45°)}$ level. Furthermore, large $E_{x(\varphi \sim 0°)}$ for feed #4 corresponds to low $XP_{(\varphi \sim 0°)}$; whereas an increase in $E_{x(\varphi \sim 45° \& 90°)}$ does not correspond to any such lowering in $XP_{(\varphi \sim 45° \& 90°)}$ values. Therefore, it is difficult to infer any definite correlation between E_x and XP.

The variation in E_y throws some meaningful information. Somewhat lower $E_{y(\varphi \sim 90°)}$ leads to lower $XP_{(\varphi \sim 90°)}$ which seems to be valid for all feed configurations. Over the D-plane, $E_{y(\varphi \sim 45°)}$ gradually increases for feeds #1–3, and also remains low for feed #4. Such variation in $E_{y(\varphi \sim 45°)}$ does not corroborate the variation in $XP_{(\varphi \sim 45°)}$ and this is true for the E-plane ($\varphi = 0°$). This analysis ensures that E_y has a predominant role on XP production over $\varphi = 90°$ plane only. But looking at the nature of those XP values, one may apprehend that, E_y would not be the sole factor!

Related observations for H_x are as follows: $H_{x(\varphi \sim 90°)}$ reveals the maximum value for feed #3 and produces the highest $XP_{(\varphi \sim 90°)}$ level. Similarly, feed #4 exhibits the lowest $H_{x(\varphi \sim 90°)}$ which corresponds to the lowest $XP_{(\varphi \sim 90°)}$ value. But $H_{x(\varphi \sim 45°)}$ and $XP_{(\varphi \sim 45°)}$ do not follow alike correlations and the reason may be one of these

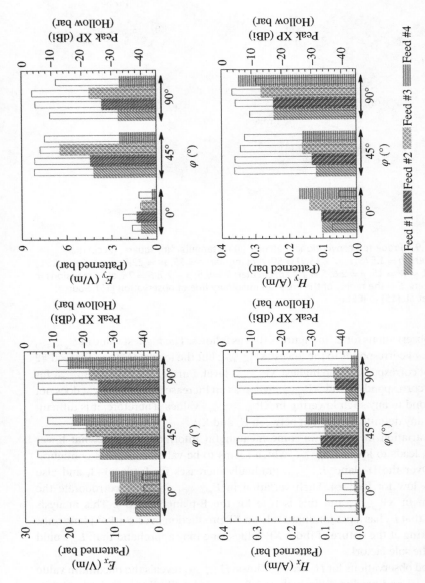

Figure 5.10 Variation in electric and magnetic fields at radial distance S and resulting XP levels at three different radiation planes for a rectangular patch. Source: Rafidul et al. [15] © IEEE. Feeds #1: coaxial probe, #2: coplanar microstrip inset feed, #3: coplanar microstrip edge feed, and #4: aperture coupled feed. Parameters as in Figure 5.9.

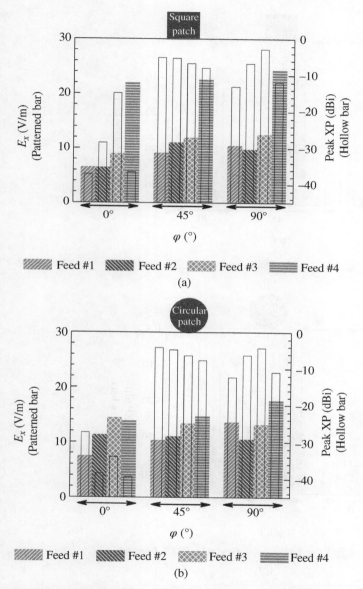

Figure 5.11 Variation in E_x at radial distance S and resulting XP levels at three different radiation planes for: (a) square patch, (b) circular patch. Source: Rafidul et al. [15] © IEEE. Feed and other as in Figure 5.10.

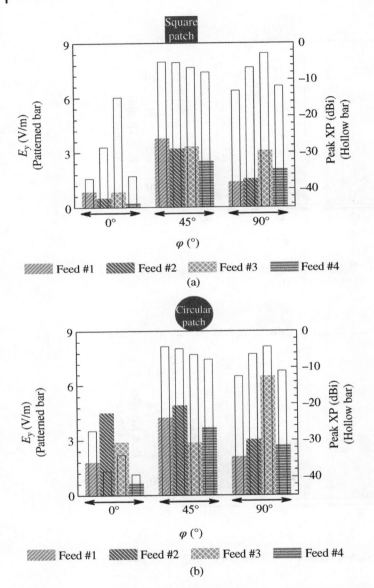

Figure 5.12 Variation in E_y at radial distance S and resulting XP levels at three different radiation planes for: (a) square patch, (b) circular patch. Source: Rafidul et al. [15] © IEEE. Feed and other as in Figure 5.10.

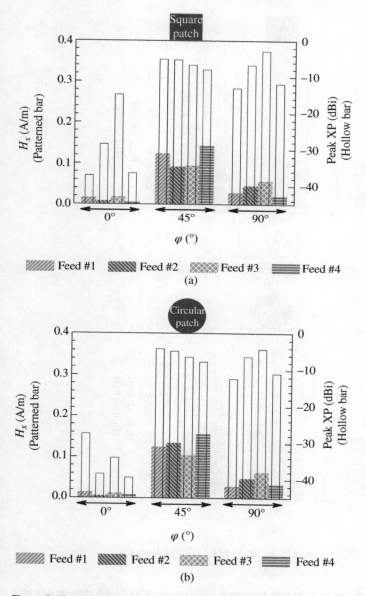

Figure 5.13 Variation in H_x at radial distance S and resulting XP levels at three different radiation planes for: (a) square patch, (b) circular patch. Source: Rafidul et al. [15] © IEEE. Feed and other as in Figure 5.10.

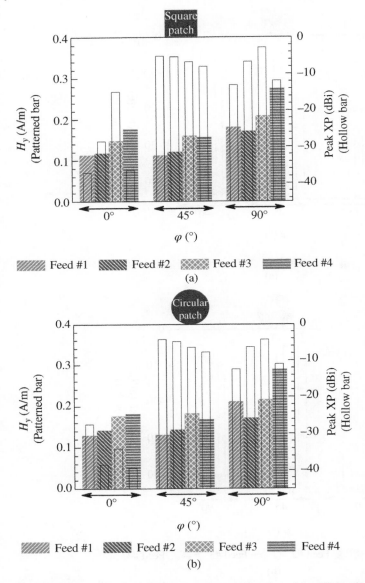

Figure 5.14 Variation in H_y at radial distance S and resulting XP levels at three different radiation planes for: (a) square patch, (b) circular patch. Source: Rafidul et al. [15] © IEEE. Feed and other as in Figure 5.10.

two possibilities: (i) $H_{x(\varphi\sim45°)}$ has no influence on $XP_{(\varphi\sim45°)}$ or (ii) $H_{x(\varphi\sim45°)}$ has a weak influence along with a stronger impact of H_y on $XP_{(\varphi\sim45°)}$. The contribution of H_y indicates two distinct characteristics: (i) $H_{y(\varphi\sim0°)}$ maintains inverse relations with $XP_{(\varphi\sim45° \& 90°)}$, and (ii) $H_{y(\varphi\sim90°)}$ reveals an insignificant relation with $XP_{(\varphi\sim90°)}$. They are valid for all feed configurations.

The above statements focusing on the correlation between H_y and XP level help in developing more insights. Feeds #1 and #2 exhibit low and closely identical $H_{y(\varphi\sim0°)}$ which correlates equally higher value of $XP_{(\varphi\sim45°)}$. The results for feed#4 establish a firm conviction to this. Feed #4 reveals maximum $H_{y(\varphi\sim0°)}$ with the lowest $XP_{(\varphi\sim45°)}$ but revealing almost no impact on $XP_{(\varphi\sim90°)}$ levels.

5.3.2 Square and Circular Patches

The above analysis executed for a rectangular patch has been validated through identical examinations with square and circular patch geometries and they are portrayed through Figures 5.11–5.14. Here also, the XP level is found to be inconclusive with respect to E_x. As in rectangular patch, $XP_{(\varphi\sim90°)}$ is directly related to $E_{y(\varphi\sim90°)}$ for both square and circular patches with a single exception. The quantity E_y in a square patch with feed #2 does not follow the trend and the reason behind such anomaly is due to a direct contribution of $H_{x(\varphi\sim90°)}$ as revealed from Figure 5.13a. Feed #2 in a square patch results in relatively larger $H_{x(\varphi\sim90°)}$ than that by feed #4 (Figure 5.13a). But $H_{y(\varphi\sim0°)}$ (Figure 5.14) follows the same trend as in a rectangular patch: larger $H_{y(\varphi\sim0°)}$ produces lower $XP_{(\varphi\sim45°)}$. The outcome of the above analysis may be summarized as [15]:

(i) A reduction in $XP_{(\varphi\sim90°)}$ is obtained by reducing $E_{y(\varphi\sim90°)}$ and/or $H_{x(\varphi\sim90°)}$.
(ii) A reduction in $XP_{(\varphi\sim45°)}$ is obtained by increasing $H_{y(\varphi\sim0°)}$.
(iii) A reduction in $XP_{(\varphi\sim45° \& 90°)}$ is obtained by increasing $H_{y(\varphi\sim0°)}$ along with a decrease in $E_{y(\varphi\sim90°)}$ and/or $H_{y(\varphi\sim90°)}$.

5.4 DGS-Based Designs for Low XP in All Radiation Planes

The above drawn inferences illustrate how a designer could reduce the XP radiations in a microstrip patch by controlling three specific source fields such as E_y, H_x, and H_y. A brief sketch of the scheme is listed in Table 5.1. This illustrates how a control of fields could be accomplished by moderating the surface currents i_x and i_y. The typical ground plane current beneath a resonant patch comprises both i_x and i_y as symbolically shown in Figure 5.15a. The design goal, according to Table 5.1, is to block i_y and facilitate i_x. The circuit theory proposes to turn the

Table 5.1 Desired surface fields for minimizing XP radiations.

Source Fields and design target	Control mechanism	Target surface impedance	Remark
$E_{y(\varphi\sim90°)}\downarrow$	By suppressing orthogonal mode	—	By resonant or non-resonant DGS
$H_{x(\varphi\sim90°)}\downarrow$	GP current $i_{y(\varphi\sim90°)}\downarrow$	$Z_{S1}\uparrow$	Engineered GP surface to enhance $i_{x(\varphi\sim0°)}$ and weaken $i_{y(\varphi\sim90°)}$
$H_{y(\varphi\sim0°)}\uparrow$	GP current $i_{x(\varphi\sim0°)}\uparrow$	$Z_{S2}\downarrow$	

$$Z_{S1} = \left|\frac{E_x}{H_y}\right|, Z_{S2} = \left|\frac{E_y}{H_x}\right|$$

2D ground plane in to a combination of x-polarized conducting strips as depicted in Figure 5.15b. Again it needs some further modifications based on the practical needs as shown in Figure 5.15c. This is meant for accommodating a feed line to the patch and holding electric fields beneath both patch and feeding line.

5.4.1 Design of Microstrip Line-Fed Circular Patch Antenna

The ground plane scheme as depicted through Figure 5.15 has been implemented for a C-band circular patch as shown in Figure 5.16 [15]. Figure 5.17 shows the photographs of a set of prototypes bearing engineered and conventional ground planes.

The modified ground indeed does not affect the input impedance of the patch and hence not the operating frequency. That is the beauty of this design as revealed from the results in Figure 5.18. The radiation characteristics across both $\varphi = 90°$ and $45°$ planes are of our interest and they are examined in Figure 5.19. The peak gain remains unaffected compared to that caused by a conventional ground plane. A remarkable reduction in H-plane XP level by 16 dB is evident. The real achievement is obtained in D-plane there the XP radiation get reduced by about 11 dB. This low XP feature is consistent over the entire operating band.

The idea behind the above design is to enhance $H_{y(\varphi\sim0°)}$ in order to suppressing $XP_{(\varphi\sim45°)}$ and diminish $H_{x(\varphi\sim90°)}$ to get rid of $XP_{(\varphi\sim90°)}$ as inferred in Table 5.1. The XP performance over both H- and D-planes is also as per the expectation. The source parameters, therefore, have been revisited in Figure 5.20 in comparison with the conventional version. The requirement of low $H_{x(\varphi\sim90°)}$ for low $XP_{(\varphi\sim90°)}$ is clearly evident. The other requirement of higher $H_{y(\varphi\sim0°)}$ for reduced $XP_{(\varphi\sim45°)}$ is also ensured. These numerical values have been extracted on an arc of radius $S = 1.25L$ (Figure 5.9) for three representative φ values. Figures 5.21 and 5.22

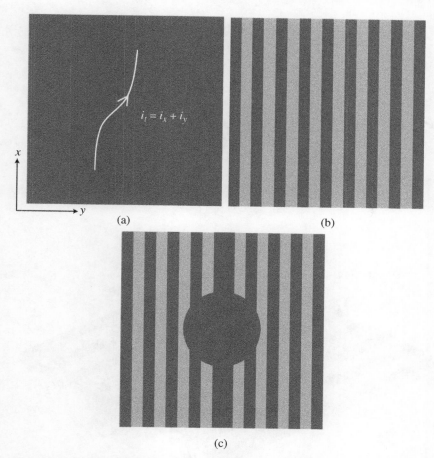

$$i_t = i_x + i_y$$

(a)

(b)

(c)

Figure 5.15 The ground plane (GP) beneath a microstrip patch and required engineering [15]: (a) typical surface current on a conventional GP; (b) conjecture for an engineered GP to enhance i_x; (c) modified geometry in order to accommodate EM fields underneath a circular patch and its feed line. Source: Rafidul et al. [15] © IEEE.

provides a scope to visualization them pictorially over the entire surface. They clearly indicate low $H_{x(\varphi \sim 90°)}$ along with high $H_{y(\varphi \sim 0°)}$ in the newly conceived structure. The change in surface impedance as portrayed in Figure 5.23 over the three strategic planes also justifies the expected variation in the source fields.

5.4.2 Design of a Coax-Fed Rectangular Patch

The approach of handling the source fields adopted in Figure 5.15 should be independent of patch geometry as well as feed configuration. The studies in

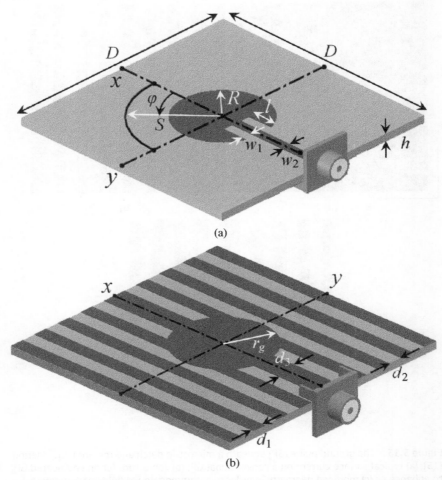

(a)

(b)

Figure 5.16 Engineered geometry based on the analytical studies [15]: (a) isometric view from top side; (b) isometric view from back side. Parameters: $R = r_g = 9.5$, $l = 6$, $w_1 = 2$, $w_2 = 2.2$, $d_1 = d_2 = 2.8$, $d_3 = 5$, $h = 1.575$, $\varepsilon_r = 2.33$, $D = 50$ (all dimensions in millimeter) [15]. Source: Rafidul et al. [15] © IEEE.

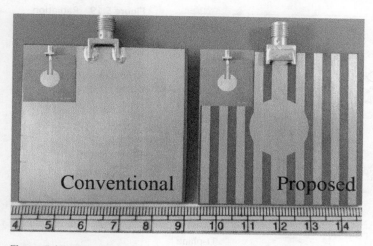

Figure 5.17 Photographs of the prototype of the antenna shown in Figure 5.16 [15]. Parameters as in Figure 5.16. Source: Rafidul et al. [15] IEEE.

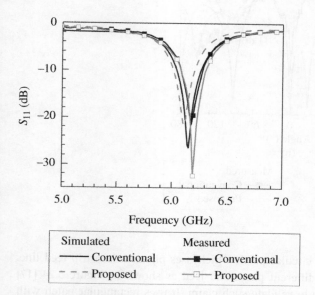

Figure 5.18 S_{11} versus frequency of the proposed prototypes shown in Figure 5.17 [15]. Source: Rafidul et al. [15] © IEEE.

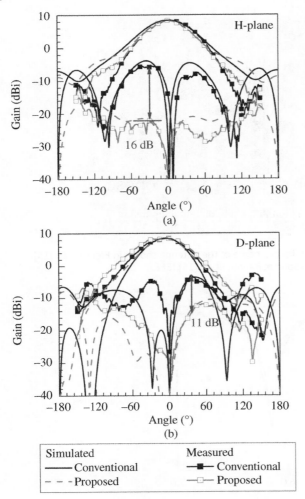

Figure 5.19 Radiation patterns of the circular patch prototypes (Figure 5.17): (a) H-plane; (b) D-plane [15]. Source: Rafidul et al. [15] © IEEE.

Section 5.4.1 deals with circular patch which uses planar microstrip feed line. Therefore, altogether a different patch and feed as shown in Figure 5.24 [17] has been examined here to validate such claim. It uses rectangular patch with coaxial probe feeding. The ground plane (GP) engineering follows the same architecture as in Figures 5.15 and 5.16 and optimized antenna is found to resonate around 3.1 GHz with both conventional and DGS-based ground. The measured radiation characteristics over H- and the D-planes are shown in Figure 5.25. Here also, no major change occurs for the co-polarized patterns, but a

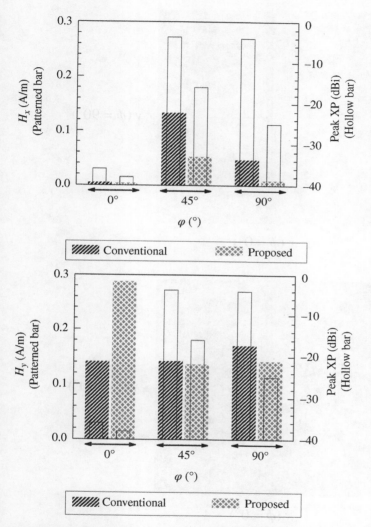

Figure 5.20 Variation in H_x and H_y at radial distance S for the antenna shown in Figure 5.16 and resulting XP levels at three different radiation planes.

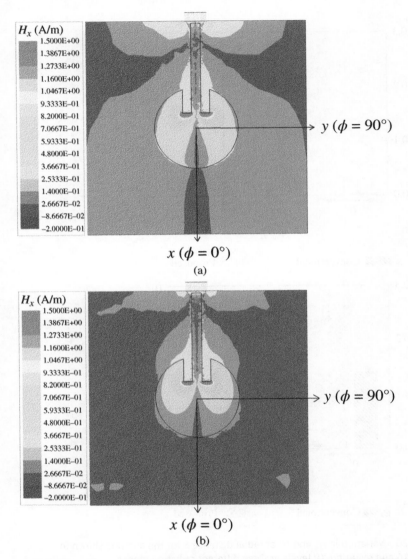

Figure 5.21 Simulated H_x field distribution on the upper surface of the substrate [15]: (a) conventional antenna; (b) DGS-based antenna. Source: Rafidul et al. [15] © IEEE.

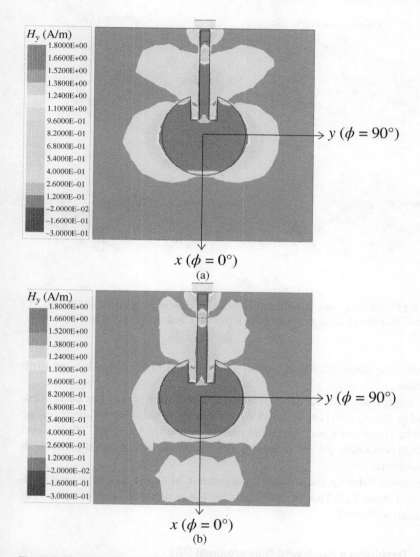

Figure 5.22 Simulated H_y field distribution on the upper surface of the substrate [15]: (a) conventional antenna; (b) DGS-based antenna. Source: Rafidul et al. [15] © IEEE.

Conventional GP   Engineered GP

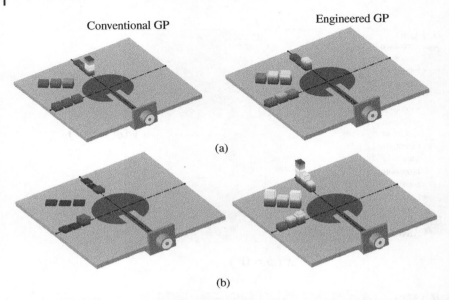

(a)

(b)

Figure 5.23 3D bar graphs showing impedance variation over a quadrant [15]: (a) $|E_x/H_y|$ with scale 0 to 290 Ω; (b) $|E_y/H_x|$ with scale 0 to 800 Ω. Source: Rafidul et al. [15] © IEEE.

noticeable reduction in XP level is observed. It is about 13 dB in H-plane and 10 dB in D-plane.

The improvement in XP performance for all skewed planes has been documented in Figure 5.26 [17]. It indicates one quarter of radiation sphere revealing successful suppression over the planes passing through $0° < \varphi \leq 90°$. The XP level with DGS peaks near $\varphi \approx 40°$ indicating 10 dB suppression compared to that with a conventional GP.

The source fields for the antenna configuration in Figure 5.24 have been compared in Figure 5.27. This ensures attaining the required field modifications as demanded by Table 5.1.

5.4.3 Designing a Patch with Non-proximal DGS

The sources of cross-polarized radiations over the entire radiation planes were addressed from a different angle in [16]. The design in [16] is a successful extension of [24] made by the same group of authors and that was actually the first attempt for skewed radiation planes. The authors [16] focused on the

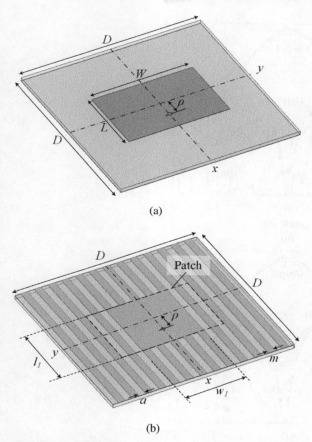

Figure 5.24 Coax-fed rectangular patch on an engineered GP [17]: (a) top view; (b) bottom view. $L = 30$ mm with $W/L = 1.5$ using RT Duroid 5870 substrate. Source: Adapted from Dutta et al. [17].

inherent asymmetry of the GP current for a conventional microstrip patch as depicted in Figure 5.28. Such asymmetry is believed to be a manifestation of XP sources [26]. Therefore, a non-proximal symmetric DGS was conceived [24] to obtain reasonable symmetry in GP current. The ground plane geometry of [24] is shown in Figure 5.29 and its impact of GP current distribution is visible in Figure 5.30. It appears much more symmetric compared to that in Figure 5.28. But no improvement in XP was achieved over other than H-plane [24]. The main point of observation is that the surface current symmetry could not help in D-plane XP scenario.

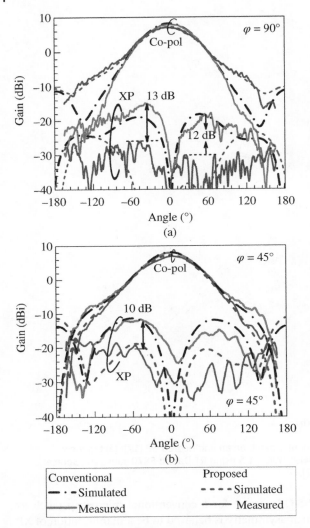

Figure 5.25 Radiation patterns of the antenna shown in Figure 5.24 compared with its conventional version: (a) H-plane ($\varphi = 90°$); (b) D-plane ($\varphi = 45°$) [17]. Source: Dutta et al. [17] © IEEE.

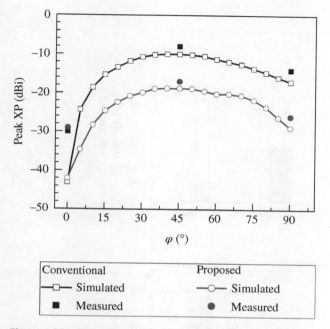

Figure 5.26 Peak XP level as a function of radiation plane φ for antenna (Figure 5.24) with and without DGS [17]. Source: Dutta et al. [17] © IEEE.

A subsequent investigation by the same group of authors [16] gave attention to improvise the DGS geometry of [24] by trial-and-error method to optimize the surface current polarization (maximize i_x and minimize i_y). The improvised geometry is shown in Figure 5.31 [16] which follows the original design (Figure 5.29 [24]) with some additional current paths. This small modification eventually results in a major change in the surface current polarization as evident from a comparison in Figure 5.32.

The additional stubs in Figure 5.32a draw more x-polarized currents compared to those in Figure 5.32b and effectively minimize y-polarized currents on the outer GP corners. The measurements with a set of prototypes shown in Figure 5.33 reveal some interesting results as furnished in Figures 5.34 and 5.35. The radiation patterns in Figure 5.34 obtained at 6.3 GHz predict at least 5 dB suppression over both H- and D-planes. In the D-plane, the asymmetric pattern may exhibit the relative reduction up to 9 dB. The XP suppression is consistent over entire radiation planes as ensured in Figure 5.35. Moreover, the order of suppression is independent of the patch aspect ratio.

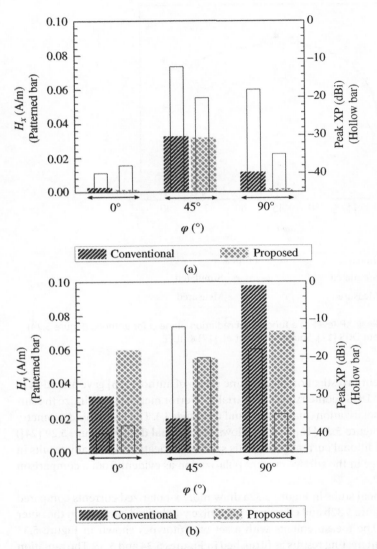

Figure 5.27 Simulated surface fields and XP radiations of the antenna shown in Figure 5.24 compared with its conventional counterpart: (a) H_x components; (b) H_y components.

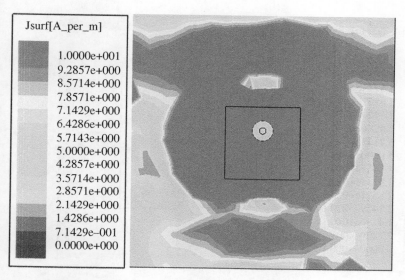

Jsurf[A_per_m]
1.0000e+001
9.2857e+000
8.5714e+000
7.8571e+000
7.1429e+000
6.4286e+000
5.7143e+000
5.0000e+000
4.2857e+000
3.5714e+000
2.8571e+000
2.1429e+000
1.4286e+000
7.1429e−001
0.0000e+000

Figure 5.28 Simulated portray of surface current for a coax-fed microstrip patch with conventional ground plane [24]. Source: Kumar et al. [24] © IEEE.

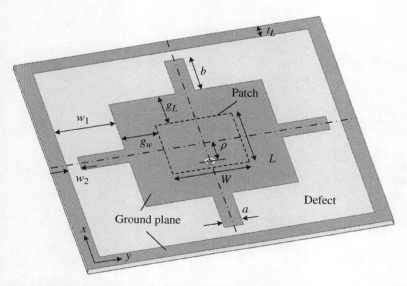

Figure 5.29 Schematic diagram of a non-proximal DGS integrated microstrip patch. Source: Adapted from Kumar et al. [24].

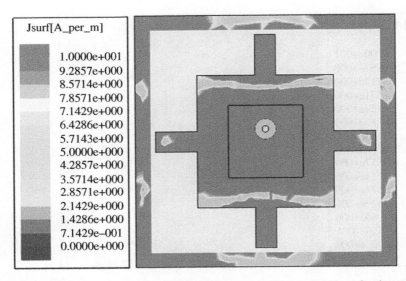

Figure 5.30 Simulated portray of surface current on the ground plane for the antenna sketched in Figure 5.29 [24]. Source: Kumar et al. [24] © IEEE.

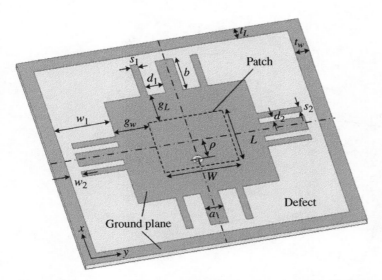

Figure 5.31 Schematic geometry of a non-proximal DGS after modification over and above the geometry shown in Figure 5.29. Source: Adapted from Pasha et al. [16].

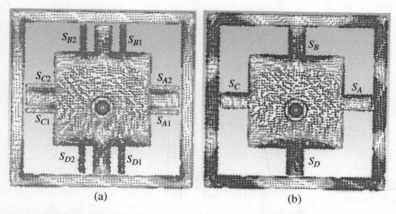

Figure 5.32 Simulated conduction currents on the non-proximal DGS surfaces:
(a) improvised geometry [16]; (b) original geometry [24]. Source: Pasha et al. [16] © IEEE.

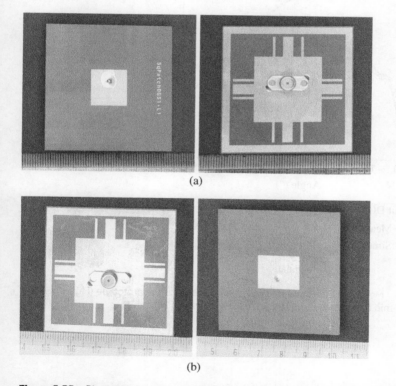

Figure 5.33 Photographs of a set of prototypes of the antenna in Figure 5.31 viewed
from the ground plane side [16]: (a) antenna with a square microstrip element;
(b) antenna with a rectangular microstrip element. Source: Pasha et al. [16] © IEEE.

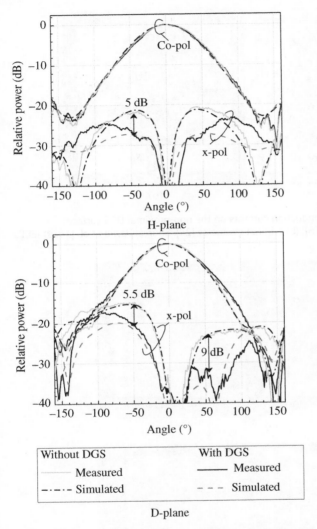

Figure 5.34 Measured and simulated radiation patterns of the prototype (Figure 5.33b) obtained at mid band frequency [16]. Source: Pasha et al. [16] © IEEE.

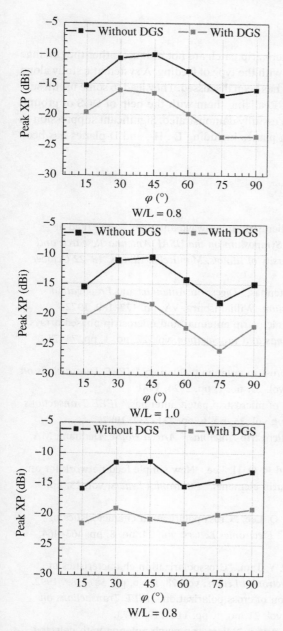

Figure 5.35 Relative change in XP level of the antenna shown in Figure 5.31 as a function of radiation plane for three specific *W/L* values [16]. Source: Pasha et al. [16] © IEEE.

5.5 Conclusion

The sources of XP fields in a microstrip patch are not unique; rather they change with their own shape as well as with the type of feeding. A systematic study along with a comprehensive overview has been discussed. They lead to some useful suggestions and design guidelines. Realizing them with the help of DGS or ground plane engineering has been successfully demonstrated. Significant suppression in XP radiations over all radiation planes including E-, H-, and D-planes has been demonstrated.

References

1 G. A. Deschamps, (1953) "Microstrip microwave antennas," presented at the *Proceedings of the Third Symposium on the USAF Antenna Research and Development Program, University of Illinois, Monticello, Illinois, 18–22 October 1953.*

2 J. Q. Howell, "Microstrip antennas," *Digest* on *Antennas* and *Propagation Society International Symposium*, Williamsburg, VA, pp. 177–180, 1972.

3 R. E. Munson, "Conformal microstrip antennas and microstrip phased arrays," *IEEE Transactions on Antennas and Propagation*, vol. 22, no. 1, pp. 74–78, 1974.

4 K. Carver and J. Mink, "Microstrip antenna technology," *IEEE Transactions on Antennas and Propagation*, vol. 29, no. 1, pp. 2–24, 1981.

5 Hansen, "Cross polarization of microstrip patch antennas," *IEEE Transactions on Antennas and Propagation*, vol. 35, no. 6, pp. 731–732, 1987.

6 I. J. Bahl and P. Bhartia, *"Microstrip Antennas,"* Artech House, Dedham, MA, 1980.

7 M. L. Oberhart, Y. T. Lo, and R. Q. H. Lee, "New simple feed network for an array module of four microstrip elements," *Electronics Letters*, vol. 23, no. 9, pp. 436–437, 1987.

8 T. Huynh, K. F. Lee, and R. Q. Lee, "Cross polarization characteristics of rectangular patch antennas," *Electronics Letters*, vol. 24, no. 8, pp. 463–464, 1988.

9 K. F. Lee, K. M. Luk, and P. Y. Tam, "Crosspolarization characteristics of circular patch antenna," *Electronics Letters*, vol. 28, no. 6, pp. 587–589, 1992.

10 A. C. Ludwig, "The definition of cross polarization," *IEEE Transactions on Antennas and Propagation*, vol. 21, no. 1, pp. 116–119, 1973.

11 D. Guha, M. Biswas, and Y. Antar, "Microstrip patch antenna with defected ground structure for cross polarization suppression," *IEEE Antennas and Wireless Propagation Letters*, vol. 4, pp. 455–458, 2005.

12 C. Kumar and D. Guha, "Higher mode discrimination in a rectangular patch: new insight leading to improved design with consistently low cross-polar radiations," *IEEE Transactions on Antennas and Propagation*, vol. 69, no. 2, pp. 708–714, 2021.

13 D. Dutta, D. Guha, and C. Kumar, "Microstrip patch with grounded spikes: a new technique to discriminate orthogonal mode for reducing cross-polarized radiations," *IEEE Transactions on Antennas and Propagation*, vol. 70, no. 3, pp. 2295–2300, 2022.

14 S. Bhardwaj and Y. Rahmat-Samii, "Revisiting the generation of cross polarization in rectangular patch antennas: a near-field approach," *IEEE Antennas and Propagation Magazine*, vol. 56, no. 1, pp. 14–38, 2014.

15 S. Rafidul, D. Guha, and C. Kumar, "Sources of cross-polarized radiation in microstrip patches: multi-parametric identification and insights for advanced engineering," *IEEE Antennas and Propagation Magazine*, 2022 (DOI: https://doi.org/10.1109/MAP.2022.3143434).

16 M. I. Pasha, C. Kumar, and D. Guha, "Mitigating high cross-polarized radiation issues over the diagonal planes of microstrip patches," *IEEE Transactions on Antennas and Propagation*, vol. 68, no. 6, pp. 4950–4954, 2020.

17 D. Dutta, S. Rafidul, D. Guha, and C. Kumar, "Suppression of cross-polarized fields of microstrip patch across all skewed and orthogonal radiation planes," *IEEE Antennas and Wireless Propagation Letters*, vol. 19, no. 1, pp. 99–103, 2020.

18 P. S. Hall and C. M. Hall, "Coplanar corporate feed effects in microstrip patch array design," *Proceedings of the Institution of Electrical Engineers*, vol. 135, no. 3, pp. 180–186, 1988.

19 W. H. Hsu and K. L. Wong, "Broad-band probe-fed patch antenna with a U-shaped ground plane for cross-polarization reduction," *IEEE Transactions on Antennas and Propagation*, vol. 50, no. 3, pp. 352–355, 2002

20 Z. N. Chen and M. Y. W. Chia, "Broad-band suspended probe-fed plate antenna with low cross-polarization level," *IEEE Transactions on Antennas and Propagation*, vol. 51, no. 2, pp. 345–346, 2003.

21 W. S. T. Rowe and R. B. Waterhouse, "Edge-fed patch antennas with reduced spurious radiation," *IEEE Transactions on Antennas and Propagation*, vol. 53, no. 5, pp. 1785–1790, 2005.

22 C. Kumar and D. Guha, "Reduction in cross-polarized radiation of microstrip patches using geometry independent resonant-type defected ground structure (DGS)," *IEEE Transactions on Antennas and Propagation*, vol. 63, no. 6, pp. 2767–2772, 2015.

23 C. Kumar and D. Guha, "Asymmetric geometry of defected ground structure for rectangular microstrip: a new approach to reduce its crosspolarized fields,"

IEEE Transactions on Antennas and Propagation, vol. 64, no. 6, pp. 2503–2506, 2016.

24 C. Kumar, M. I. Pasha, and D. Guha, "Microstrip patch with nonproximal symmetric defected ground structure (DGS) for improved cross-polarization properties over principal radiation planes," *IEEE Antennas and Wireless Propagation Letters*, vol. 14, pp. 1412–1414, 2015.

25 High Frequency Structure Simulator (HFSS) v.12. (2012). ANSYS, Pittsburgh, PA, USA.

26 C. Kumar and D. Guha, "Defected ground structure (DGS)-integrated rectangular microstrip patch for improved polarization purity with wide impedance bandwidth," *IET Microwaves, Antennas, and Propagation*, vol. 8, no. 8, pp. 589–596, 2014.

6

DGS-Based Low Cross-Pol Array Design and Applications

6.1 Introduction

The most of the antennas discussed so far are standalone microstrip elements. But many operational systems need array of elements for higher gain, scanning beam, or beam shaping purpose. Such array designs have evolved from the early days in various approaches [1–3]. A DGS integrated array would aim improved performance, but it demands serious attention to overcoming the main constraint of adequate space on the ground surface. Such a deployment issue is absent in a standalone configuration. Moreover, the feed network in an array multiplies the space complexities along with an additional concern of spurious radiations from that. As such, it is next to impossible to replicate the performance in a DGS integrated array that is achievable in a standalone antenna with DGS.

Hence, the primary challenge is to realize physically viable simple and compact DGS for printed arrays. Some innovations in geometry as well as concept of sharing could help in addressing the challenge. Various applications include reduction in mutual coupling and suppression of cross-polarized (XP) fields along with handling adverse effects on the array radiation patterns. All of these aspects have been covered based on the available information.

6.2 Low Cross-Pol Microstrip Array Design

Controlling cross-polar radiation from microstrip antennas has been one of the primary areas of DGS integrated designs. The aperture-fed arrays are known for their low XP behavior over the principal planes. But, probe- and microstrip-fed designs are prone to generating higher XP radiations even in arrays. Hence, microstrip arrays with these two types of feeding are of our primary interest and discussed in this section. For a large 2D array, a 2 × 2 sub-array could be considered as the basic

Defected Ground Structure (DGS) Based Antennas: Design Physics, Engineering, and Applications,
First Edition. Debatosh Guha, Chandrakanta Kumar, and Sujoy Biswas.

building block, and therefore, the DGS integrated designs have been validated for 2×2 geometries bearing different feeding structures.

6.2.1 Coax-Fed Microstrip Array

Historically, DGS started its journey for mitigating XP issues in 2005 with a coax-fed circular microstrip patch [4]. Subsequently, several DGS shapes befitting with varying patch geometries have been reported. One of them is shown in Figure 6.1a [5], which suits well with varying aspect ratio of rectangular patches. This is also discussed in Section 4.4.3. A pair of DGS units is required to serve a single patch. But in a compact array, there may not be enough space to accommodate more than one slot in between two adjacent elements.

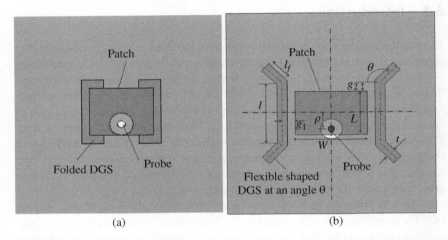

(a)   (b)

Figure 6.1 Schematic view of a DGS-integrated probe-fed rectangular patch: (a) layout with "[" and "]" type folded-DGS and (b) folded arm being a function of "θ" [6].

An advanced variant of Figure 6.1a is shown in Figure 6.1b along with their appreciable performance in Figure 6.2. The orientation of the folded arm determined by "θ" does not play much significant role in its performance [6]. Thus, it offers an advantageous feature particularly when multiple patch elements are deployed in sequence as depicted in Figure 6.3a. Figure 6.3b,c depicts a prototype of such 2×2 sub-array that uses $\theta_1 = 180°$ and $\theta_2 = 0°$ with element spacing, $s = 0.5\lambda_0$. The DGS layout ensures each antenna element to be supported by symmetric DGSs, although the pattern may vary from element to element.

Indeed the flexibility in DGS shape enables it serving two adjacent microstrip elements without compromising in performance [6]. It would be interesting to note that each H-plane row bearing n elements needs only $(n + 1)$ DGS units instead of $2n$. Figure 6.4 shows the measured radiation patterns of the array of

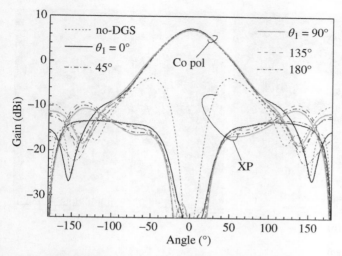

Figure 6.2 Simulated H-plane radiation patterns of the antenna in Figure 6.1 (b) with different fold angle θ. Parameters: $L = 8.6$, $W/L = 1.6$, $\rho = 3.1$, $g_1 = g_2 = 0$, $l_f = 3.13$, $l = 10.1$, $t = 1.5$, Ground plane size 60×60; all dimensions are in mm, substrate thickness 1.575 mm, $\varepsilon_r = 2.33$ [6]. Source: Kumar et al. [6] © IEEE.

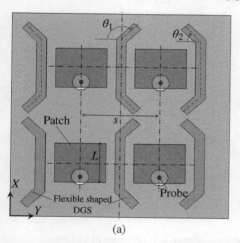

(a)

Figure 6.3 A 2×2 array of rectangular microstrip patch sharing a flexibly shaped DGS between the elements: (a) schematic view from the ground plane side, (b) photograph of a prototype – top view, and (c) photograph viewed from the back side [6]. Parameters: $L = 8.5$, $W/L = 1.3$, $\rho = 2.2$, $g_1 = 1.975$, $g_2 = 0.75$, $l = 11.5$, $l_f = 1.525$, $s = 15$, $\theta_1 = 180°$, $\theta_2 = 0°$, all dimensions are in mm and other parameters as shown in Figure 6.2. Source: Kumar et al. [6] © IEEE.

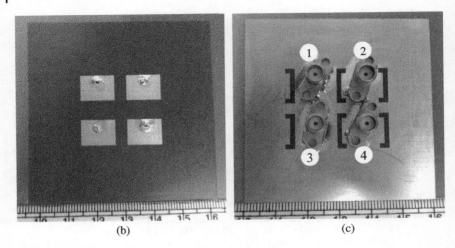

Figure 6.3 (*Continued*)

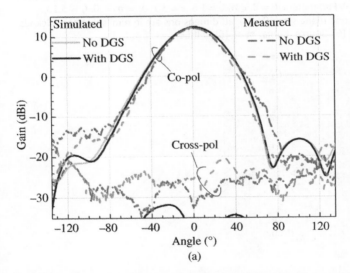

Figure 6.4 Radiation patterns of the prototype (Figure 6.3) obtained at 10 GHz:
(a) E-plane and (b) H-plane. Parameters as in Figure 6.3 [6]. Source: Kumar et al. [6]
© IEEE.

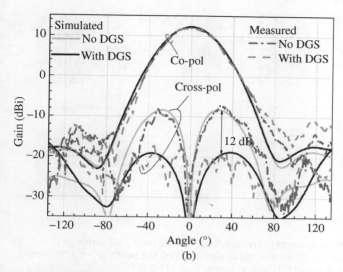

Figure 6.4 (*Continued*)

Figure 6.3, which is excited by signals bearing identical phase and amplitude. As in the case of a standalone antenna, the array also reveals no change in the co-polarized radiation patterns in the presence of the DGS. The measured peak gain is about 12.2 dBi. However, the peak XP radiations in the H-plane indicate about 12 dB suppression that leads to more than 30 dB co- to cross-polar isolation over the principal planes, which is truly significant for practical applications.

6.2.2 Microstrip Line-Fed Array

As discussed in Chapters 4 and 5, the source of the XP radiations significantly varies from feed to feed. Hence, the feature occurring in a microstrip line-fed rectangular patch differs from that occurring in a probe-fed element. The planar feed lines are exposed to the same radiation space as that of the patch, and hence any radiations from the discontinuities like steps, bends, junction, or stubs contribute to the XP fields in addition to the natural causes. The order of that contribution significantly increases in arrays where feed network incorporates multiple bends, steps, and junctions as shown in Figure 6.5a. This issue was addressed partially in ref. [7], and a simplified feed technique shown in Figure 6.5b evolved.

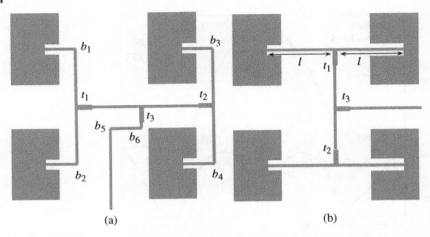

(a)   (b)

Figure 6.5 Schematic diagram of microstrip feed network for a 2×2 array: (a) conventional approach and (b) improvised design avoiding the bends ($b_1 - b_6$). Sections $t_1 - t_3$: impedance transformers [7]. Source: Kumar et al. [7] © IEEE.

Two different types of patches, one with nominal inset feed and the other with deep inset feed have been used [7]. Such modification helps in avoiding six bends ($b_1 - b_6$) in a 2×2 configuration. Figure 6.6 shows a design with simplified feed along with its simulated radiations that promise considerable improvement in the XP scenario. Minimization of spurious radiations from the feeds helps in reducing the XP level by about 20 dB even over E-plane. This performance has been considered as a reference for further exploration toward higher XP isolation.

One may refer to Figure 4.27 that depicts the need of "composite DGS" for a standalone microstrip-fed patch – one pair to address the orthogonal mode and the other pair to check spurious radiations from the feed line [8]. But hardly any room could be afforded to accommodate composite DGS in the array environment as shown in Figure 6.6a. It is equally impossible to deploy DGS at each point of feed discontinuity. Hence, a tradeoff has been conceived as sketched in Figure 6.7a. It exploits minimum possible DGS units serving both patch and feed at a time [9].

Another practical aspect needs to be considered by an antenna engineer. During simulation, a discrete port [10] is used to excite the feed line and the results are shown in Figure 6.6b,c. But in practice, the port is an SubMiniature version A (SMA) connector and that SMA to microstrip junction contributes considerably to the XP level. This reality ideally demands yet another set of DGS around this junction, but no adequate room is available.

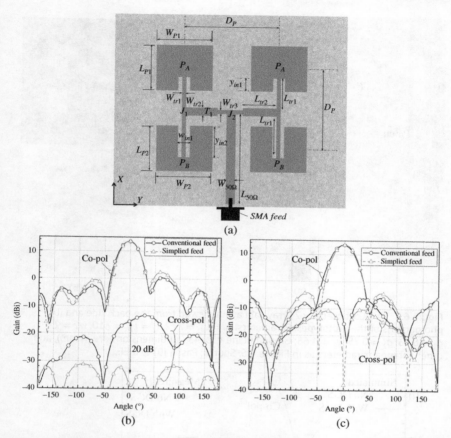

Figure 6.6 A typical 2×2 array and its simulated radiation patterns: (a) array layout, (b) E-plane patterns, and (c) H-plane patterns. Parameters: $L_{P1} = 7.72$; $L_{P2} = 7.10$, $W_{P1} = 9.27$, $W_{P2} = 8.53$, $w_{in1} = 1.5$, $y_{in1} = 2.18$, $y_{in2} = 5.70$, $D_P = 17.5$, $W_{tr1} = 0.57$, $W_{tr2} = 1.24$, $W_{tr3} = 0.57$, $W_{50\Omega} = 2.23$, $L_{tr1} = 8.37$, $L_{tr2} = 5.37$, $L_{50\Omega} = 17.21$; Substrate CuClad-250 with $\varepsilon_r = 2.55$ and thickness $= 0.787$, ground plane size: 40×40; all dimensions are in mm [7]. Source: Kumar et al. [7] © IEEE.

The DGS configuration in Figure 6.7a takes care of its scalability for larger array design without any further change. Figure 6.7b shows the photographs of the realized hardware and its measured radiation patterns are depicted in Figure 6.8. The co-polarized patterns with and without DGS look very similar. The measured peak gain at resonance is about 14 dBi. The E-plane XP values are also comparable for both the configurations, whereas the H-plane XP level reduces by about 7 dB.

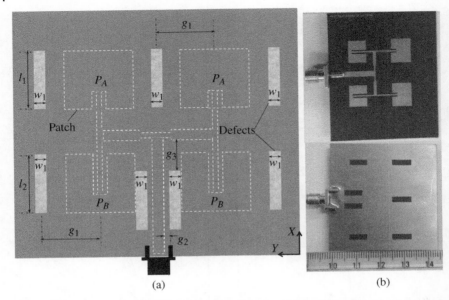

(a)   (b)

Figure 6.7 A DGS-based 2×2 array. (a) Layout viewed from the back side and (b) photographs of the prototypes. Parameters: $l_1 = L_{P1} = 7.72$, $l_2 = L_{P2} = 7.10$, $w_1 = 2.5$, $g_1 = 8.73$, $g_2 = 0.75$, $g_3 = 4.66$; Ground plane 45×45 (all dimensions are in mm), and other antenna parameters as in Figure 6.6. Source: Pasha [9] © IEEE.

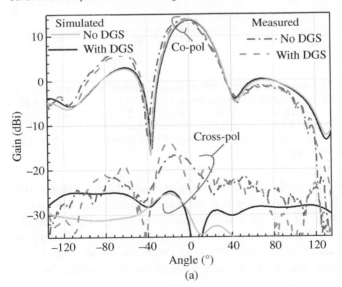

(a)

Figure 6.8 Radiation patterns of the microstrip-fed 2×2 patch array (Figure 6.7) with and without DGS ($f = 11.65$ GHz): (a) E-plane and (b) H-plane. Parameters as in Figure 6.7 [9].

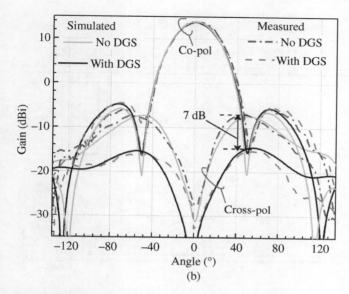

Figure 6.8 (*Continued*)

A practical question may arise when an SMA connector is used as the feeding port. Should it be printed circuit board (PCB) mount or flange mount type? These two different ways mounting are shown in Figure 6.9, and they are individually examined with a 50 Ω X-band transmission line. The nature of the propagating fields is compared in Figure 6.10 for which the simulated portrays are captured with identical scale of field intensity. The flange mount SMA (Figure 6.9a) reveals uniform field distribution compared to that due to the PCB mount configuration (Figure 6.9b). The scattering parameter S_{21} examined in Figure 6.11 also ensures its superiority over the entire band. Hence, flange-mount technique is recommended to excite a microstrip feed for practical antennas. But at the same time, relatively

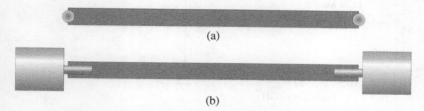

Figure 6.9 Two different types of SMA mounts connected at two ends of a microstrip line: (a) flange mount configuration; (b) PCB mount configuration.

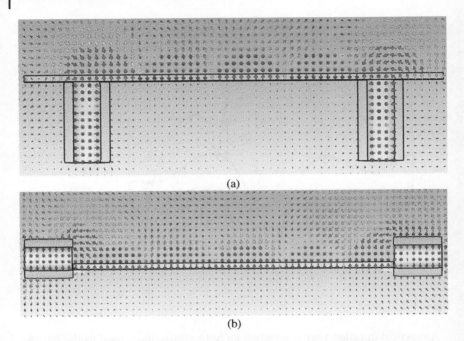

Figure 6.10 Substrate electric fields portray at 11.5 GHz for microstrip lines shown in Figure 6.9: (a) flange mount; (b) PCB mount.

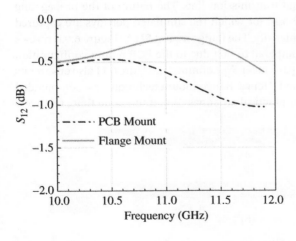

Figure 6.11 Simulated S_{21} characteristics compared for the microstrip lines examined in Figures 6.9 and 6.10.

wider lateral spreading of substrate fields appears as a characteristic feature of flange-mount case. This recommends strategic DGSs on its both sides to address any spurious fields and help minimizing such spurious reductions.

Keeping this in view, the low XP single element design in Section 5.3.1 has been extended to 2×2 array [11] as shown in Figure 6.12. The strip like DGS restricts

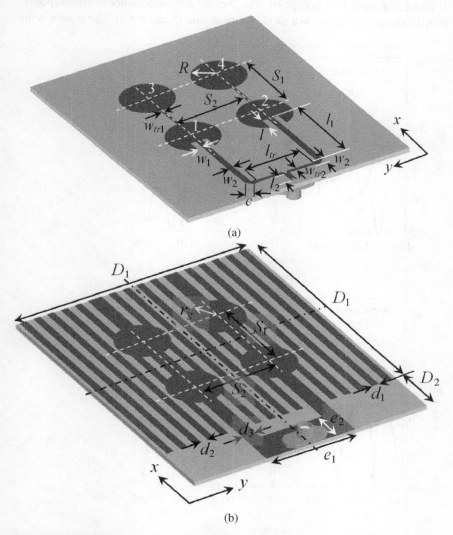

(a)

(b)

Figure 6.12 DGS-based 2×2 array [11]: (a) isometric view from top side; (b) isometric view from the back side. Parameters as in [11]. Source: Rafidul et al. [11] © IEEE.

the feed line orientation and hence a series-fed configuration has been used. The array offers more than 13.5 dBi peak gain, which is marginally higher by 0.8 dB compared to that achievable by a conventional ground plane (GP). The radiation patterns are shown in Figure 6.13, which significantly bears the evidence of XP reduction over both H- and D-planes. As much as 16 dB reduction is promised in H-plane and about 5 dB in D-plane. The D-plane XP performance in array is poorer by 6 dB compared to that in a single element case (Section 5.3). The reason is the

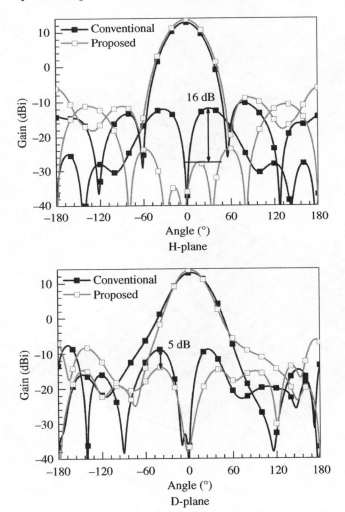

Figure 6.13 Simulated H-and D-plane patterns of 2×2 array depicted in Figure 6.12 [11]. Source: Rafidul et al. [11] © IEEE.

Table 6.1 2 × 2 array with different Feed and DGS configurations

Array feeding type	DGS type	XP Suppression in dB		Comments
		H-plane	D-plane	
Probe feed [6]	Flexible-shaped DGS	12	Does not work	XP performance not uniform over all radiation planes
Corporate feed with PCB mount SMA [9]	Slot DGS	7	Does not work	
Corporate feed with flange mount SMA (Figure 6.21) [11]	Strip like DGS	16	5	Truly low XP solution, working over all radiation planes.

x-polarized surface current that loses the linear tracks in array due to intermediate circular bases on the GP (Figure 6.12). Table 6.1 provides a comparative study when the DGS-based array is fed by different mechanisms.

6.3 Array Design for Reduced Mutual Coupling

The potential of DGS in reducing mutual coupling between two adjacent microstrip elements sharing the common substrate and ground plane was realized at the very early phase of developing DGS technology [12]. That investigation actually characterized concentric ring DGS, and in that connection, it demonstrated suppression of mutual coupling in a 2 × 1 array of circular patches. The mutual coupling indeed is an unwanted phenomenon that severely degrades the operation of an array. Like microstrip, dielectric resonator antenna (DRA) array also shares a common ground for the array elements, and a similar technique was found equally effective in a DRA array [13]. This technique drew attention of a large section of microstrip engineers as revealed from its rapid growth and various practical applications. Comprehensive information on this topic has been provided in Chapter 7, which is exclusively dedicated to the DGS-based engineering in reducing mutual coupling effects.

6.4 DGS-Based Array for Different Applications

The basic potentials of DGS were reported at the very beginning of its inception in 2005–2006 in the form of reduced XP characteristics in planar antenna [4] and suppressed mutual coupling between two adjacent antenna elements [12]. Both have found their applications in planar arrays especially in the airborne systems since the technique does not add any additional volume or weight. A wide range of applications has been discussed here.

6.4.1 Elimination of Scan Blindness

The scan blindness is an inherent feature of a planar-phased array that restricts the scan range. The main reason is a large input mismatch caused by a coincidence of the propagation constant of Floquet mode with that of the natural substrate mode. This unwanted characteristic of a phased array was addressed with the help of DGS technique [14, 15] by reducing the mutual coupling, and thus it firmly established the scope of DGS-based array for eliminating the Scan Blindness. This topic has been elaborately discussed under the purview of Chapter 7.

6.4.2 Millimeter-Wave Imaging with Suppressed XP

As in communication system, high cross-pol isolation is a critical requirement for the imaging systems at microwave or millimeter wave frequencies. The co-polarized signal scattered from an object bears the signature of its smooth parts whereas the orthogonally oriented signal contains the information of its edges. Hence, a combined processing of these two signals provides the complete impression, and hence they need to be processed independently. No crosstalk

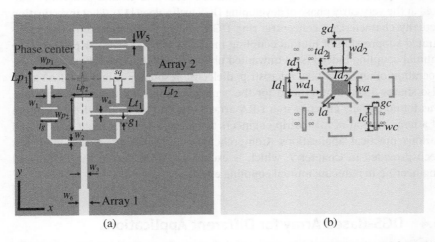

(a)   (b)

Figure 6.14 Two 2×1 arrays with mutually orthogonal orientation: (a) viewed from the top side and (b) viewed from the back side indicating DGS layout [16]. Source: Rezaei et al. [16] © IEEE.

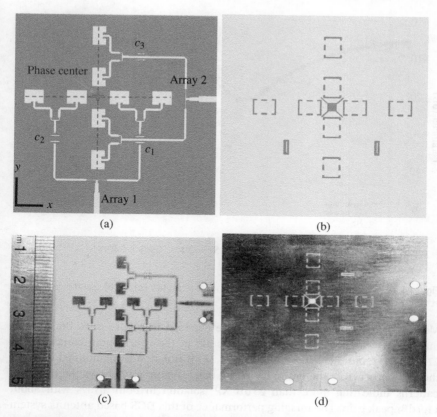

Figure 6.15 4×1 arrays in vertical and horizontal orientations: (a) top view, (b) bottom view with DGS layout, (c) photograph of a prototype – top view, and (d) photograph from the ground plane side [16]. Source: Rezaei et al. [16] © IEEE.

between these two mutually orthogonal signals is acceptable and as such each set of antennas has to be free from any XP radiations [16]. This, therefore, becomes a very special and critical requirement. The actual system demands more than 10 dBi gain, and thus the antenna has to be configured as arrays. A design was developed in ref. [16] as depicted through Figures 6.14 and 6.15. A pair of 2×1 arrays as shown in Figure 6.14a maintains mutually orthogonal polarization to pick up signals of each individual polarization. Both arrays have a common phase center marked as "+," and their DGS layout is shown in Figure 6.14b. This eventually results in more than 25 dB co- to XP isolation along with 9.6 dBi gain.

The 2×1 array depicted in Figure 6.14 has been extended to 4×1 array as shown in Figure 6.15 to realize the actual operating system. The cross-overs in the feed are carefully designed to maintain the phase and amplitude uniformity.

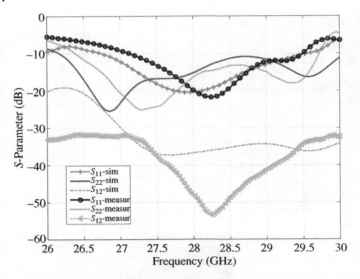

Figure 6.16 *S*-parameters of 4 × 1 array as shown in Figure 6.15 [16]. Source: Rezaei et al. [16] © IEEE.

Measured *S*-parameters of this array are depicted in Figure 6.16. Both arrays marked as "Array 1" and "Array 2" operate in the overlapping bandwidth of 27–29.5 GHz and more importantly, over 32 dB isolation is maintained between the two ports across the entire operating band. Figure 6.17 shows the radiation patterns indicating more than 25 dB XP isolation in both arrays with about 11.9 dBi peak gain. The imaging performance of this DGS-based antenna system exhibits improved resolution of the image boundary [16].

6.4.3 High-Performance Rectenna Array

Microstrip antennas have become popular in wireless power transfer that enables obtaining direct current (DC) energy from radio frequency (RF) radiations. They are commonly called rectifying antenna or simply *rectenna*. Small rectennas are generally used in terrestrial Internet of Things (IoT) type applications. Supplying power to a CubeSat through a rectenna is one of such applications where the RF energy is imparted upon the CubeSat by a nearby satellite located at higher orbit [17]. A scheme of the rectenna array is shown in Figure 6.18, which bears circularly polarized elements for consistent reception of the energy irrespective of the orientation of the transmitting or receiving systems.

The design for reduced RADAR cross-section (RCS) is shown in Figure 6.18 [17]. This uses clover-shaped DGS in combination with complementary split ring resonator (CSRR) sharing the common ground plane. The DGS acts as a tank circuit,

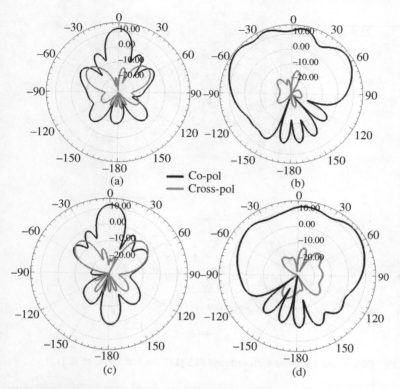

Figure 6.17 Simulated radiation patterns of the arrays shown in Figure 6.15 ($f = 28$ GHz): (a) H-plane: Array I, (b) E-plane: Array I, (c) H-plane: Array II, and (d) E-plane: Array II [16]. Source: Rezaei et al. [16] © IEEE.

and its varying reactance with frequency leads to the diversification of current distribution and the associated scattered wave [17]. Additionally, the negative permittivity of CSRR changes the phase of the scattered fields and the combined effect results in reduced RCS. Figure 6.19 shows the measured results of the DGS-based rectenna array, which are self-explanatory. It promises about 5 dB RCS reduction over 5.6–6.0 GHz operating frequencies.

6.4.4 Enhancement of Scanning Range

The DGS has been explored in improving scanning range of a metasurface array [18]. The antenna schematic is shown in Figure 6.20. The top layer comprises metasurface as the radiating element, which is fed by an array of nine apertures. The middle layer is the common ground bearing the feeding apertures with the feed lines beneath them. This configuration is referred to as array I [18]. This

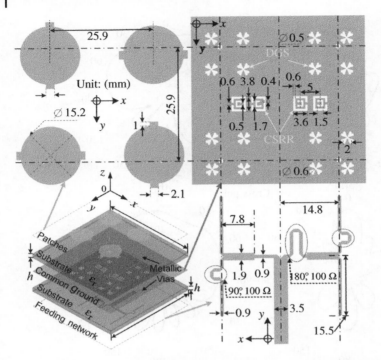

Figure 6.18 DGS-based Rectenna with reduced RCS [17]. Source: Wang et al. [17] © IEEE.

antenna normally scans up to 45° with 3 dB drop in gain relative to the boresight peak value. For boresight radiation, all apertures are uniformly excited with identical phase and amplitude resulting in considerably good gain and low XP level. But for scanned beam, the excitation phases change and they disturb the current distribution on the radiating elements as shown by the simulated surface current portrays in Figure 6.21a. This leads to degradation in radiation pattern [18].

The cross components of the surface current have been addressed by introducing a set of meander-line-shaped DGSs on the common ground plane. This configuration is referred to as array II. The DGS acts like a polarizer and does the necessary corrections in polarization of surface currents. It can be visualized from the simulated portrays in Figure 6.21, and its impact is manifested through the radiation characteristics depicted through Figures 6.21b and 6.22. The radiation due to array II appears to be more symmetric around the boresight. The scan beam performance in Figure 6.22 indicating its scanning ability up to 50° with nominal scan loss of 3 dB. On the other hand, it results in only 1.2 dB scan loss at 40° scan angle whereas an identical case without DGS reveals a scan loss by about 2.4 dB [18].

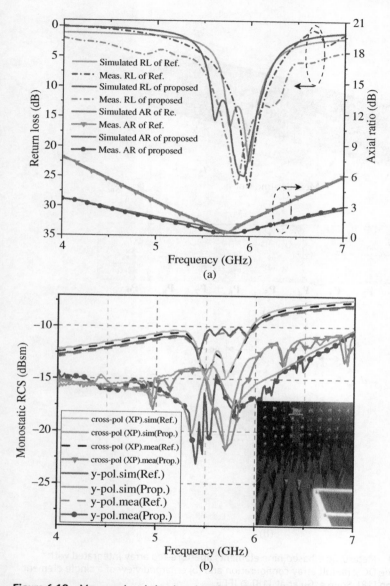

Figure 6.19 Measured and simulated performances of the antennas with and without DGS: (a) return loss and axial ratio versus frequency and (b) RCS versus frequency [17]. Source: Wang et al. [17] © IEEE.

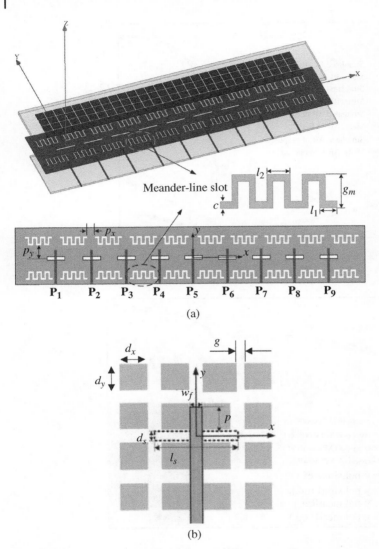

Figure 6.20 Metasurface-based nine elements aperture-fed array integrated with meander-line DGS: (a) full array configuration and (b) enlarged view of a single element configuration [18]. Source: Gu et al. [18] © IEEE.

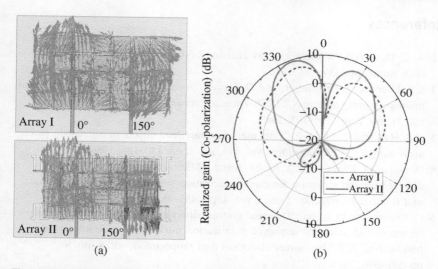

Figure 6.21 Comparison of the array performance (Figure 6.20) with and without DGS: (a) simulated portray of surface current and (b) H-plane radiation patterns [18]. Source: Gu et al. [18] © IEEE.

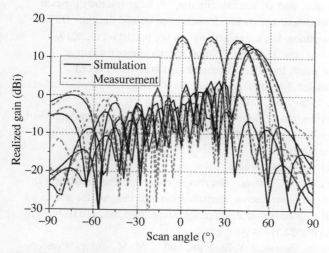

Figure 6.22 Measured and simulated radiation patterns of DGS-based array II (Figure 6.20) as a function of scan angle [18]. Source: Gu et al. [18] © IEEE.

References

1 A. G. Derneryd, "Microstrip array antenna," *6th European Microwave Conference*, Rome, pp. 339–343, 1976.

2 J. Ashkenazy, P. Perlmutter, and D. Treves, "A modular approach for design of microstrip array antennas," *IEEE Transactions on Antennas and Propagation*, vol. 31, no. 1, pp. 190–193, 1983.

3 P. S. Hall and C. M. Hall, "Coplanar corporate feed effects in microstrip patch array design," *IEE Proceedings*, vol. 135, no. 3, pp. 180–186, 1988.

4 D. Guha, M. Biswas, and Y. M. M. Antar, "Microstrip patch antenna with defected ground structure for cross polarization suppression," *IEEE Antennas and Wireless Propagation Letters*, vol. 4, pp. 455–458, 2005.

5 C. Kumar and D. Guha, "Defected ground structure (DGS)-integrated rectangular microstrip patch for improved polarization purity with wide impedance bandwidth," *IET Microwaves, Antennas and Propagation*, vol. 8, no. 8, pp. 589–596, 2014.

6 C. Kumar, M. I. Pasha, and D. Guha, "Defected ground structure integrated microstrip array antenna for improved radiation properties," *IEEE Antennas and Wireless Propagation Letters*, vol. 16, pp. 310–312, 2017.

7 C. Kumar, V. S. Kumar, and D. Venkataramana, "A large microstrip patch array with a simplified feed network: a low cross-polarized design," *IEEE Antennas and Propagation Magazine*, vol. 61, no 05, pp. 105–111, 2019.

8 M. I. Pasha, C. Kumar, and D. Guha, "Simultaneous compensation of microstrip feed and patch by defected ground structure for reduced cross-polarized radiation," *IEEE Transactions on Antennas and Propagation*, vol. 66, no. 12, pp. 7348–7352, 2018.

9 M. I. Pasha, "Novel designs of defected ground structure-integrated microstrip antennas and arrays for improved radiation characteristics," PhD Thesis, University of Calcutta, 2019.

10 CST Simulation Software, CST Studio, 2016.

11 S. Rafidul, D. Guha, and C. Kumar, "Sources of cross-polarized radiation in microstrip patches: multi-parametric identification and insights for advanced engineering," *IEEE Antennas and Propagation Magazine*, 2022 (DOI: https://doi.org/10.1109/MAP.2022.3143434).

12 D. Guha, S. Biswas, M. Biswas, J. Y. Siddiqui, and Y. M. M. Antar, "Concentric ring-shaped defected ground structures for microstrip applications," *IEEE Antennas and Wireless Propagation Letters*, vol. 5, pp. 402–405, 2006.

13 D. Guha, S. Biswas, T. Joseph, and M. T. Sebastian, "Defected ground structure to reduce mutual coupling between cylindrical dielectric resonator antennas," *Electronics Letters*, vol. 44, no. 14, pp. 836–837, 2008.

14 H. Moghadas, A. Tavakoli, and M. Salehi, "Elimination of scan blindness in microstrip scanning array antennas using defected ground structure," *AEÜ - International Journal of Electronics and Communications*, vol. 62, no. 2, pp. 155–158, 2008.

15 D.-B. Hou, S. Xiao, B.-Z. Wang, L. Jiang, J. Wang, and W. Hong, "Elimination of scan blindness with compact defected ground structures in microstrip phased array," *IET Microwaves, Antennas and Propagation*, vol. 3, no. 2, pp. 269–275, 2009.

16 M. Rezaei, H. Zamani, M. Fakharzadeh, and M. Memarian, "Quality improvement of millimeter-wave imaging systems using optimized dual polarized arrays," *IEEE Transactions on Antennas and Propagation*, vol. 69, no. 10, pp. 6848–6856, 2021.

17 S. C. Wang, M. J. Li, and M. S. Tong, "A high-performance rectenna for wireless power transfer in CubeSat," *IEEE Antennas and Wireless Propagation Letters*, vol. 19, no. 12, pp. 2197–2200, 2020.

18 L. Gu, Y. W. Zhao, Q. M. Cai, Z. P. Zhang, B. H. Xu, and Z. P. Nie, "Scanning enhanced low-profile broadband phased array with radiator-sharing approach and defected ground structures," *IEEE Transactions on Antennas and Propagation*, vol. 65, no. 11, pp. 5846–5854, 2017.

7

DGS Based Mutual Coupling Reduction: Microstrip Array, 5G/MIMO, and Millimeter Wave Applications

7.1 Introduction

The microstrip technology is now driven by the requirements of packing more and more components within a limited space but with high performance. This requirement is more stringent for the 5G multiple-input-multiple-output (MIMO) antennas. A microstrip antenna primarily excites radiating fields propagating through the free space as a function of $1/r$, where r is a distance from the radiator. Parallelly, a resonating patch produces surface waves that propagate through the grounded substrate and decays as $1/\sqrt{r}$. The surface wave transforms to radiating fields at the air–dielectric interface with decaying factor $1/r^2$. But it is not too serious in the case of employing sufficiently thin substrate for which the surface-to-radiating wave transition occurs at a considerably large distance [1]. A schematic diagram in Figure 7.1 helps in visualizing the effect of surface waves when a common substrate is shared by two or more radiating patches. The RF energy fed to one element

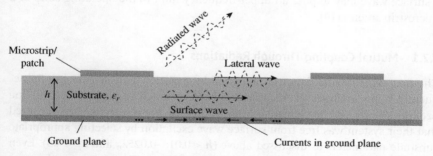

Figure 7.1 Schematic diagram indicating various channels of mutual coupling between two microstrip patches sharing a common grounded substrate.

Defected Ground Structure (DGS) Based Antennas: Design Physics, Engineering, and Applications,
First Edition. Debatosh Guha, Chandrakanta Kumar, and Sujoy Biswas.
© 2023 The Institute of Electrical and Electronics Engineers, Inc. Published 2023 by John Wiley & Sons, Inc.

gets coupled to its neighboring ones, and this phenomenon is known as mutual coupling. There might be more than one mechanism behind this.

It has both positive and negative impacts. In a few applications such as enhancing antenna bandwidth by parasitically coupled elements or reducing side lobe level in an array by excitation taper near the edge [2, 3], mutual coupling helps. But in a majority of the cases, the mutual coupling is detrimental to the antenna array performance. It reduces the gain, squints the beam, increases the side-lobe level (SLL), destroys desired nulls, and generates grating lobes along with scan blindness. In view of MIMO antennas, mutual coupling has been a burning issue denying multiple closely packed elements in one device to work. As a result, exploring several techniques to mitigate the mutual coupling has been an active area of research since the early 1990s [4]. Several possibilities and solutions have been explored over the period. The possibility of defected ground structure (DGS) in handling this issue was first reported in ref. [5]. Subsequent developments along with some fundamental aspects have been discussed in this chapter.

7.2 Mutual Coupling Mechanisms

Three specific mechanisms have been known so far. The mutual coupling occurs through (i) the process of radiation, (ii) propagation of the surface wave, and (iii) common ground plane current. The coupling through radiations is mainly determined by the radiation patterns of the individual elements and their intermediate physical separation. But that occurring through surface waves is determined by the substrate parameters like relative permittivity, ε_r, and thickness, h. Typically, a substrate with $h < 0.015$–$0.025\lambda_0$ and $\varepsilon_r < 2.5$ is not prone to surface wave excitation [2, 3, 6–8], and hence the effect of surface wave can be safely ignored. Indeed, a surface wave may impose an upper frequency limit to the operating band of a microstrip antenna [9].

7.2.1 Mutual Coupling Through Radiations

The first experimental data were reported by Jedlicka et al. in 1981 [3]. Subsequently, many analyses [8, 10, 11] have been executed to predict the coupling between the microstrip elements. In those studies, the researchers first ensured that their system was free from surface wave excitation by selecting appropriate substrate parameters as discussed above ($h < 0.015$–$0.025\lambda_0$ and $\varepsilon_r < 2.5$). Even then, a significant order of mutual coupling was detected by comparing the measured data with the analytical values [8].

7.2.2 Mutual Coupling by Surface Waves

The surface wave prominently exists when the substrate permittivity and thickness are considerably large. An experimental investigation [12] estimated its effect on the radiation patterns of microstrip patches printed on $0.02\lambda_0 - 0.03\lambda_0$ thick substrate with relative $\varepsilon_r = 10$–13. The array configuration and the radiation patterns of one of the elements are shown in Figure 7.2 [12]. A signature of edge diffraction caused by the surface waves is apparent from the high side lobes in E-plane (Figure 7.2b). But no such aberration is seen in H-plane pattern (Figure 6.2c) since coupling through surface wave modes is minimally possible across H-plane [12]. The E-plane pattern improves significantly when the ground plane edges are covered by RF absorbers (Figure 6.2d). This, indeed, prevents the surface wave from diffracting at the ground plane edges.

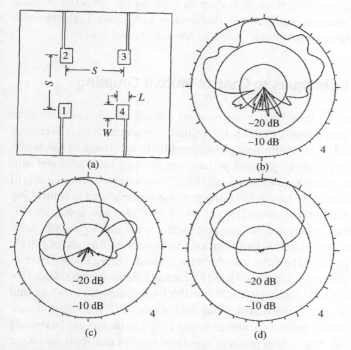

Figure 7.2 Effect of surface wave on the radiation of an individual element (# 4) in a planar array printed on a substrate with $\varepsilon_r = 13$ and $h = 0.03\lambda_0$ [12]: (a) array layout; (b) E-plane pattern; (c) H-plane pattern; (d) E-plane pattern incorporating absorbers near edges; ground plane size: $1.97\ \lambda_0 \times 1.97\ \lambda_0$. Source: Schabuert and Yngvesson [12] © IEEE.

Some interesting information is also available from the S-parameter studies [12]. The maximum coupling occurs at resonance especially across E-plane ($\varepsilon_r = 10.2$), and this remains sufficiently high even at off resonance. That study [12] reveals that the mutual coupling is a function of the physical separation between the patches and varies almost linearly with that.

7.2.3 Coupling Through Ground Plane Currents

On a microstrip platform, any antenna bears significant surface currents on its ground plane. For a common ground bearing two or more close by antenna elements, there is every possibility of mutual interactions through the surface currents. Thus, the signal fed to one element gets coupled to the other one. But it is difficult to study the nature of such coupling exclusively. However, as a rule of thumb, the mutual coupling through surface current reduces with the increase in the element-to-element separation. It, therefore, is a challenge to the antenna engineers to realize a compact design using a common ground plane.

7.3 Known Techniques to Control Mutual Coupling

The antenna engineers have been trying to avoid mutual coupling in microstrip arrays in multiple ways since 1990s [13–15]. Initially, the attempts were to control the specific modes [13] or synthesizing low permittivity substrates by micromachining techniques [14] or by physical perforations through the traditional substrates [15]. Use of electromagnetic band-gap (EBG) structures was reported in [16] and later on extended in [17]. Such EBG structure was also explored in controlling mutual coupling through radiations [18, 19]. Their basic idea was to employ vertical resonant shields in between two adjacent elements in order to trap the fields and let the neighboring elements free from any perturbations. The use of EBG is shown in Figure 7.3 [20], where the pattern on the ground (Figure 7.3a) comprises complementary split-ring resonators (SRRs) with an additional connecting slot for each unit. A pair of 5 GHz patches use 1.27-mm thick substrate with $\varepsilon_r = 3.48$, and hence surface wave effect may be ruled out. Still about 10 dB reduction in coupling was determined from the S-parameter values [20]. The impact can be visually realized from Figure 7.3b, which portrays simulated current distributions on the common ground plane with and without EBG.

EBG structure actually gave birth to the concept of DGS as narrated in Chapter 1. The DGS was also found equally effective in controlling the ground plane currents and minimizing mutual coupling between two coplanar microstrip elements [5]. A detailed account of DGS-based approaches has been addressed in the following section.

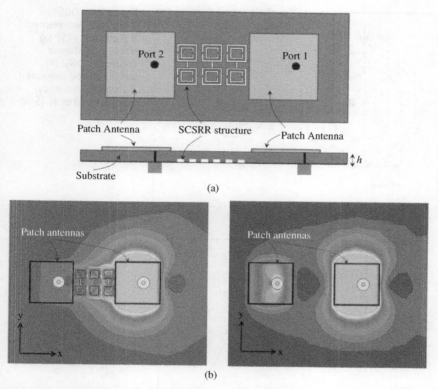

Figure 7.3 Periodic complementary split-ring resonator (SRR)-based EBG on the ground plane to reduce mutual coupling between microstrip patches [20]: (a) top and cross-sectional views; (b) Simulated surface currents with and without EBG. Source: Bait-Suwailam et al. [20] © IEEE.

7.4 DGS Based Solutions to Mutual Coupling

The ability of DGS in controlling the ground plane current and hence the mutual coupling between two microstrip elements was first reported in [5]. Like EBG-based design [17], DGS does not require additional fabrication or volume. Figure 7.4 shows an elementary scheme for two-element X-band microstrip patches [5]. One annular ring DGS concentric to either element serves the purpose. The substrate with $\varepsilon_r = 2.33$ and $h \approx 0.05\lambda$ results in a reduction by 4 dB in S_{21}. This idea was experimentally verified in [21] and was quickly picked up by different groups [22, 23]. It included dumbbell [22] and many other user friendly shapes. Their principle of operation is suppressing the surface wave, if it persists

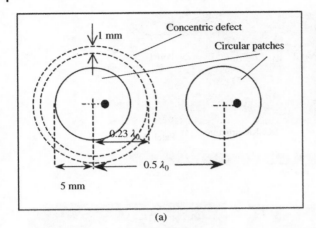

Figure 7.4 DGS-integrated two-element E-plane array [5]: (a) single ring-shaped concentric DGS; (b) measured and simulated S_{21} versus frequency. Source: Guha et al. [5] © IEEE.

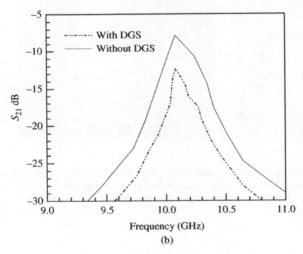

at that frequency of operation, and also controlling the common ground plane currents.

Apart from printed antennas, the DGS-based decoupling has also been explored for other radiating structures like planar inverted-F antenna (PIFA) [24, 25] and dielectric resonator antenna (DRA) [26]. Figure 7.5 shows a pair of closely packed PIFA elements side by side [24]. Their common ground plane has been serrated by multiple DGS slits and that facilitates in reducing S_{21} value by more than 5 dB.

This concept is perfectly working to decouple two S-band DRA elements aligned to their E-plane [26]. The geometry [26] looks very similar to that in Figure 7.4. Each cylindrical DRA is fed by a coaxial probe and is deployed on a thin grounded

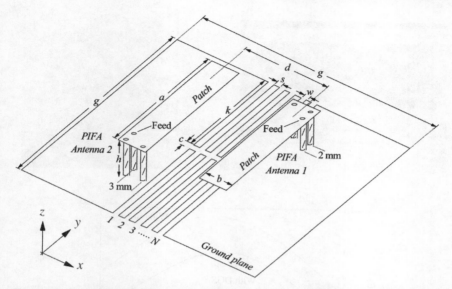

Figure 7.5 Closely packed two PIFA elements with an intermediate serrated slot DGS for suppressing mutual coupling. Source: Chiu et al. [24] © IEEE.

substrate. The measured data ensure a consistent reduction in S_{21} by about 5 dB from 3.1 to 3.4 GHz operating band. Surface wave is not a relevant term for this antenna configuration, and hence the entire decoupling is based on mitigating the ground plane current by DGS.

For the decoupling purpose, a DGS is typically deployed in between two radiating antennas. This means that the DGS is always isolated from the radiating element and continuously interacting with the common ground current and the surface waves if it persists in that case. Such a physical situation has been modeled in Section 2.6 for theoretical analysis [27]. This analysis was employed in [28] to characterize a DGS geometry for the first time. It examines two adjacent DRAs as shown in Figure 7.6a [28]. Their effectiveness in controlling the mutual coupling in the presence of an arc-DGS is clearly understandable from Figure 7.6b,c. The impact of 5 dB reduction in S_{21} is easily reflected in the simulated surface current scenario in Figure 7.6c.

Subsequent investigations have introduced several new concepts and innovative approaches. Strategic orientations of slots had created new DGS shapes [29] revealing an excellent feature of obstructing the current flow from one patch to the other. A variant of that S-shaped DGS [29] was reported in the form of a fractal as shown in Figure 7.7 [30]. It uses 3.81 mm thick Rogers thermoset microwave material (TMM) 6 substrate to design a pair of S-band patches and more than 30 dB

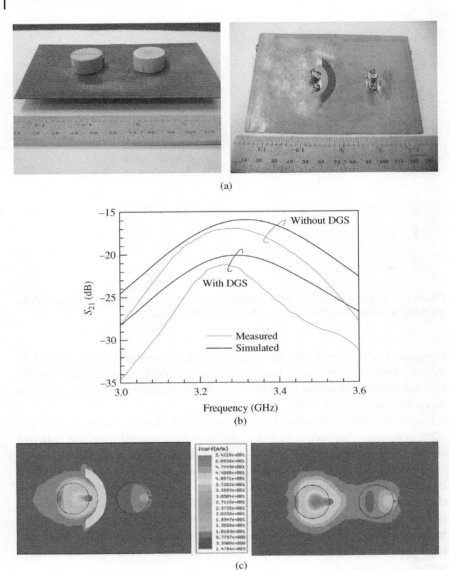

Figure 7.6 Arc-shaped DGS to reduce mutual coupling between a pair of cylindrical DRAs [28]: (a) top and back side views of a prototype; (b) measured S_{21} over the operating band; (c) Simulated surface currents with and without DGS. DRA: radius = height = 10 mm, $\varepsilon_r = 10$; substrate: $\varepsilon_r = 2.2$, $h = 0.508$ mm. Source: Biswas and Guha [28] Elsevier.

Figure 7.7 Fractal DGS to mitigate the mutual coupling between two S-band square patches etched on 3.81 mm thick Rogers TMM 6 substrate [30]: (a) second iterative fractal DGS; (b) third iterative fractal DGS. Source: Wei et al. [30] © IEEE.

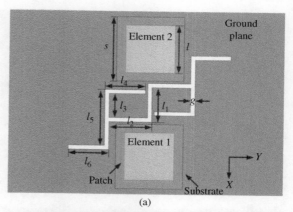

(a)

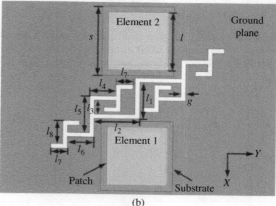

(b)

reduction in mutual coupling has been documented. Both second and third iterative fractals examined in [30] are depicted in Figure 7.7. The third iterative fractal appears to be more effective in handling the mutual coupling compared to the second iterative geometry as easily visualized from Figure 7.8. This helps in increasing the antenna efficiency along with a decrease in envelope correlation which in turn indicates its suitability for MIMO applications.

Preferred shaping of DGS correlates preferred current paths in between two elements and that is the key in choosing the DGS shape. Meandered line loading inside a defect was reported in [31] for a dual band printed PIFA design. A defect with meandered line shape has recently been conceived from the angle of common/differential mode theory [32]. The geometry is shown in Figure 7.9a,b. Two square patches are separated by a longitudinal slot coupled with a pair of meandered line shaped DGS. It uses 1.524 mm thick FR4 substrate with $\varepsilon_r = 4.4$. The ground plane current for the common mode (CM) and difference mode (DM)

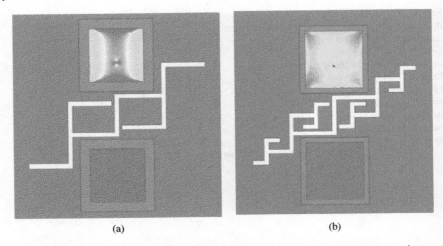

(a)   (b)

Figure 7.8 Simulated conduction currents on the patch surface of the antennas shown in Figure 7.7 in presence of (a) second iterative fractal DGS (2.525 GHz) and (b) third iterative fractal DGS (2.3 GHz). Source: Wei et al. [30] © IEEE.

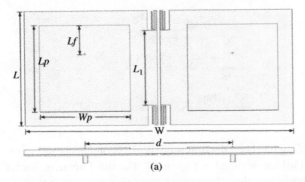

(a)

Figure 7.9 Microstrip patches with a DGS that works on common mode/difference mode approach [32]: (a) Antenna layout; (b) enlarged view of meandered line-shaped defect; (c) simulated current distributions showing the common mode (LHS) and difference mode (RHS) cases. Source: Qian et al. [32] © IEEE.

excitations is shown in Figure 7.9c [32]. The impedance characteristics of both CM and DM mutually overlap which according to common and differential mode theory ensures perfect decoupling over the matching bandwidth.

This common/difference mode analysis has also been applied to a 2×2 microstrip array [33]. This also uses machine learning technique for optimizing the DGS geometry. The specialty is in shaping the DGS in the form of "H" as shown in Figure 7.10a. The deployment is strategic to achieve decoupling in both

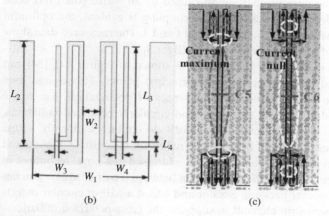

(b)

(c)

Figure 7.9 (Continued)

Figure 7.10 A 2 × 2 microstrip array with specially configured H-shaped DGS [33]: (a) schematic diagram; (b) conduction current on patch surfaces in presence of the DGS when element 1 is active. Source: Qian et al. [33] © IEEE.

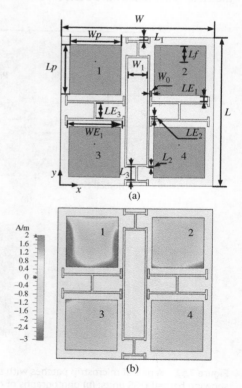

(a)

(b)

E- and H-planes. The current distribution caused by an active patch has been portrayed in Figure 7.10b [33]. Sufficient decoupling is evident, the optimum being between elements 1 and 3, and elements 1 and 4. The measured data show improvement by 27 dB for S_{13} whereas just 20 dB for S_{12}.

Arrays of defects surrounding the patches as shown in Figure 7.11 [34] can take care of mutual coupling across both principal planes. This design uses dual feed option to realize three different senses of polarization, e.g. single linearly polarized, dual linearly polarized, and circularly polarized radiations. The measured data promised more than 35 dB decoupling uniformly across both principal planes. A tradeoff between the reduction in mutual coupling and suppression of cross-polar radiations has been addressed in a novel technique in ref. [35]. That C-band design achieved isolation between a pair of square patches by introducing a coplanar parasitic isolator and added a pair of circular defects underneath each microstrip element to suppress the cross-polarized radiations. Over 19 dB isolation at resonance along with and more than 13 dB suppression in the cross-polar fields have been experimentally documented [35].

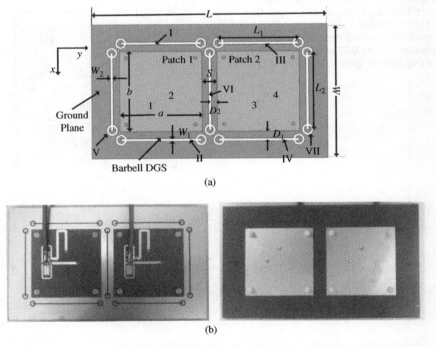

Figure 7.11 A pair of microstrip patches with slot-array DGS [34]: (a) schematic diagram showing barbell DGS units; (b) photographs of the prototype indicating the dual-feed configurations. Source: Gao et al. [34] IEEE.

7.5 Major Applications

7.5.1 Elimination of Scan Blindness in Large Arrays

Scan blindness is the most practical problem in large scanning arrays [7, 10]. The mutual coupling is the main culprit that turns the active reflection coefficient to unity for a few specific scan angles and causes radiation nulls. Controls of mutual coupling are the only remedy. The idea of using DGS to decouple two array elements originated in 2006 [5, 22] and was nurtured for applications. Within a couple

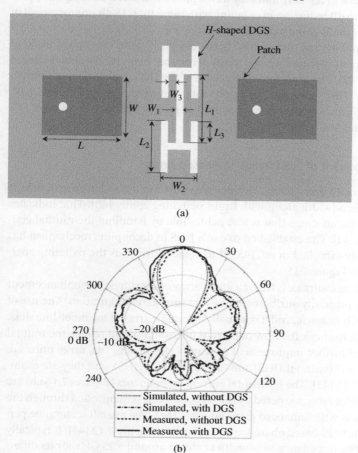

Figure 7.12 H-shaped DGS to improve scan blindness of a 2-element microstrip array [37]: (a) Layout of a C-band E-plane array; (b) E-plane patterns of fully excited array with and without DGS. Source: Hou et al. [37] with permission of Institution of Engineering and Technology.

of years, two groups of researchers proposed reduction in scan blindness for the first time [36, 37]. The results in [37] for a pair of square patches decoupled by H-shaped DGS (Figure 7.12) appear more convincing. The H-shaped DGS is more compact in size, but its higher "Q" results in steeper rejection band. An infinite phased array integrated with H-shaped DGSs was also studied in ref. [37] indicating removal of sharp nulls in E-plane patterns occurring near 58°. Other geometries like Dumbbell- [36] and "U"-shaped [38] DGSs were also reported to work successfully. The width of the stopband was enhanced by cascading two U-shaped DGSs back-to-back in ref. [38], and they were applied to a 6 GHz 2 × 1 square patch array revealing 15 dB reduction in coupling with more than 15 dB improvement in radiation null in the E-plane pattern.

A WLAN 16 element array has been designed with the possibility of element-to-element decoupling by 36 dB with the help of dumbbell-shaped split-ring DGS [39]. The array and the scheme of the DGS deployment are shown in Figure 7.13. The simulated radiation data, depicted through Figure 7.14, promise a significant improvement in radiation null by about 7–13 dB near 15°, 5–6 dB at 30°, and 10 dB at 60°.

7.5.2 Enhancement of Scan Range in Phased Array

The modern phased array demands improved performances that focus on high gain and wide bandwidth along with beam switching agility [40]. That indicates enhancement of scan range that is also achievable by handling the mutual coupling issues [41, 42]. The established role of a DGS in decoupling mechanism has been successfully exploited in ref. [43]. It uses metasurface as the radiating aperture as shown in Figure 7.15.

The radiating metasurface is fed by a linear array of apertures. The enhancement of scan range is primarily attributed to the metasurface configuration. The use of DGS is very much strategic, and it is actually a periodic array of meander-line slots. Their purpose is to check the flow of current horizontally and reduce the mutual coupling toward further improvement in scan performance. Two array units are termed as "array I" (without DGS) and "array II" (with DGS), and they are examined in Figure 7.16 [43]. The horizontal current components (Figures 7.16a,b) are found to be considerably restricted by the DGS, and this is manifested through the radiation patterns with enhanced gain (Figure 7.16c). The overall scan range performance of this DGS-based phased array is shown in Figure 7.17 [43]. It typically operates over 28% impedance bandwidth centered around 5.25 GHz for its different input ports (Figure 7.17a). The measured radiations at five specific scan angles with reference to the boresight are shown in Figure 7.17b, and this is achieved by progressive phase with uniform weightage of the inputs. The measurement shows

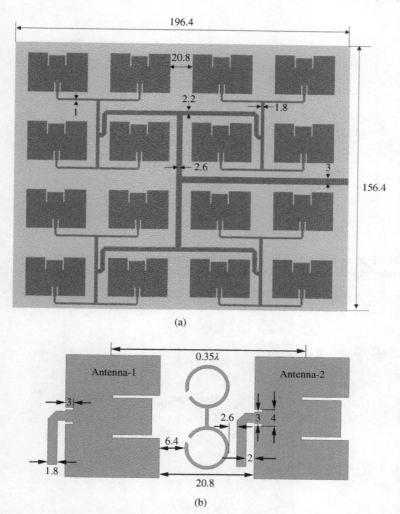

(a)

(b)

Figure 7.13 A 4×4 element DGS-integrated array investigated for reduced scan blindness [39]: (a) the array geometry bearing E-shaped patch elements; (b) dumbbell-shaped split-ring structure used as DGS [39] in between two E-shaped patched across E-plane. Source: Ghosh et al. [39] The Electromagnetics Academy.

excellent agreement with the simulated predictions. In particular, the effect of the DGS can be understood from the comparison in Table 7.1 [43]. Its impact is significantly realized at higher values of scan angle. It indeed results in an improved flat gain over the scan range.

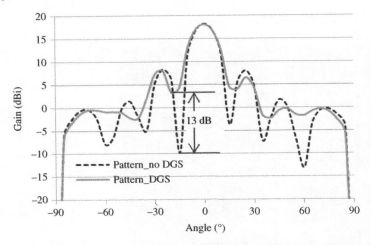

Figure 7.14 E-plane radiation patterns of the 16-element array shown in Figure 7.13 with and without DGS. Source: Ghosh et al. [39] The Electromagnetics Academy.

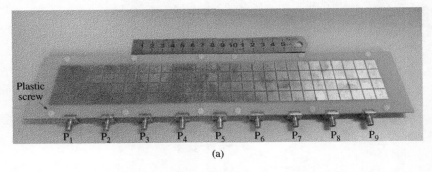

(a)

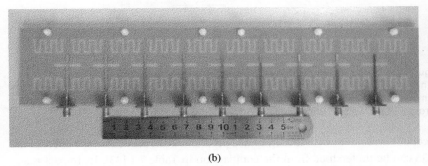

(b)

Figure 7.15 A DGS-based metasurface array [43]: (a) top view of the layer of a prototype; (b) second layer of the prototype bearing feeding apertures and meander line DGSs. Source: Gu et al. [43] IEEE.

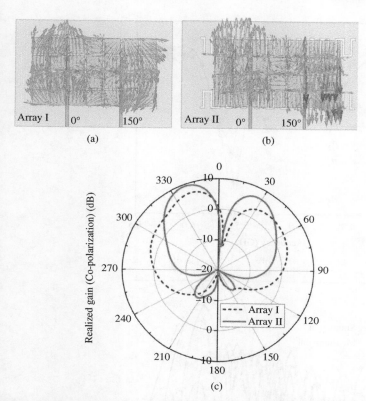

Figure 7.16 A representative study indicating the impact of DGS on a metasurface antenna unit using simulated results [43]: (a) surface current on metasurface array I without DGS; (b) surface current on metasurface array II with DGS; (c) H-plane co-polar radiation patterns for both arrays with and without DGS obtained at 5.25 GHz for 150° mutual phase difference. Source: Gu et al. [43] © IEEE.

7.5.3 DGS Based Compact Antennas for 5G/MIMO/Millimeter Wave Applications

The demand for high data rate in 5G communication systems is being addressed by UWB and/or MIMO antennas. They operate over a wide range of frequencies covering industrial, scientific, and medical (ISM) to the millimeter wave bands. The MIMO technique uses the advantages of multiple colocated antennas to increase channel capacity and making the complex RF frontend design reliable. Accommodating multiple antennas maintaining the required electrical isolation in a limited space is a big challenge for the portable devices. Defected ground or DGS could be of immense help as was realized in a few initial designs [44–46]. Figure 7.18 shows

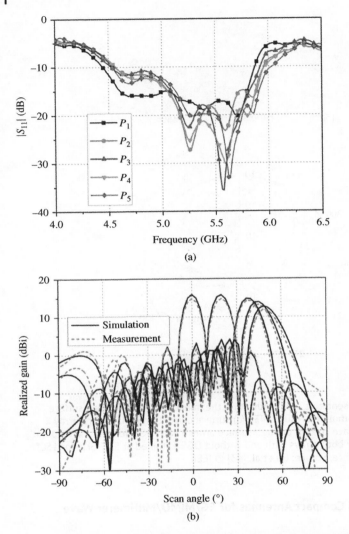

Figure 7.17 Measured data for the prototype shown in Figure 7.15 [43]: (a) S_{11} characteristics for individual feed activation; (b) H-plane scan performance obtained at 5.2 GHz. Source: Gu et al. [43] © IEEE.

a representative example [44] where a pair of sinusoid printed monopoles are decoupled by a fractal DGS. Some variants and advancements have been reported in refs. [45, 46]. All of them should maintain the required criterion that Envelope Correlation Coefficient (ECC) is less than 0.5. The ECC performance metric is expressed as [47]:

Table 7.1 The metasurface array performance with and without DGS.

Scan angle (°)	Predicted Gain at 5.25 GHz (dBi)		Remarks on gain performance
	No DGS	With DGS	
0	16.24	15.82	Decreases by 0.42–0.32 dB
10	16.15	15.80	
20	15.81	15.73	More or less comparable
30	15.10	15.43	Improved by 0.33–1.59 dB
40	13.85	14.65	
50	11.21	12.80	

Source: Gu et al. [43] IEEE.

Figure 7.18 Schematic diagram of a microstrip-fed sinusoid monopole antenna with fractal DGS placed in between them. Source: Kakoyiannis and Constantinou [44] © IEEE.

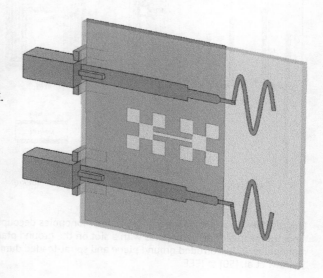

$$\rho = \frac{\left|S_{11}^*S_{21} + S_{12}^*S_{22}\right|}{\left(1 - |S_{11}|^2 - |S_{21}|^2\right)\left(1 - |S_{22}|^2 - |S_{12}|^2\right)} \tag{7.1}$$

The DGS-based MIMO antennas grew rapidly since the late 2000s [48]. Multiple simple slits were used on the ground of a four-port system [48]. Figure 7.19a depicts a simple configuration that was used in ref. [49]. An interesting and a bit complicated DGS for MIMO antennas is shown in Figure 7.19b [50]. Here, the

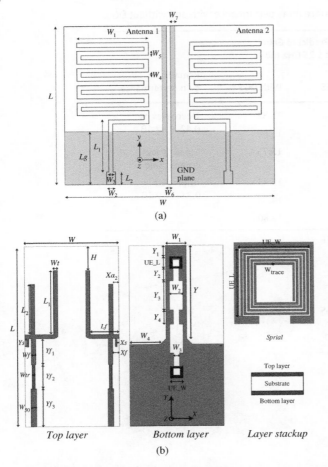

Figure 7.19 Schematic diagrams of printed monopoles decoupled by different DGSs: (a) a pair of meander line monopoles with a slot on the ground plane [49]; (b) Four shaped monopoles with protruded ground plane and spiral-loaded dumbbell DGS. Source: Sharawi et al. [50] © IEEE.

DGS looks like a specially shaped dumbbell whose two ends are loaded with spiral defects. The approach of slotted ground became popular in MIMO antennas [48–51]. Numerous investigations have been reported over the years [52–66] and a comparative study on those has been presented in Table 7.2.

The design in ref. [52] claims improved isolation and ECC compared to those in ref. [51]. A work [53] reported a very compact geometry as shown in Figure 7.20. Unlike the monopoles shown in Figure 7.19, this design [53] (Figure 7.20) achieves high isolation by intelligently deploying antenna elements perpendicular to each

Table 7.2 Comparative study of the design and performance parameters of various MIMO antennas integrated with DGS for mutual coupling reduction.

References	Dimension	Operational band	Port-to-port isolation $(S_{ij}, i \neq j)$	Antenna type	Type of DGS	No. of ports	ECC	Gain/ Efficiency	Applications
[44]	$38 \times 38\,mm^2$	2–3 GHz	−10 to −15 dB	Printed sinusoidal monopole	Fractal DGS	Dual	<0.05	—	Wireless sensor nodes
[45]	$40 \times 45\,mm^2$	2–3 GHz	−15 dB	Printed quarter wave monopole	Two H-shaped DGS in dumbbell configuration	Dual	—	—	MIMO antenna
[46]	$30 \times 65\,mm^2$	2.26–2.5 GHz 5.41–5.79 GHz	Less than −25 dB in both bands	Folded monopole	Two slots	Dual	<0.1	2.37 dBi @ 2.4 GHz 4.34 dBi @ 5.6 GHz	Dual band MIMO
[48]	$37 \times 28\,mm^2$	2.4–2.5 GHz	Less than −25 dB between any two port	Two square ring patch and two slot	Multiple slits	Four	<0.022	2.84 dBi (Square ring) 3.52 dBi (Slot)	MIMO antenna
[49]	$40 \times 50\,mm^2$	760–886 MHz	−12 dB	Meander Line antenna	Single and multiple slots	Dual	<0.3	2.2 dBi	MIMO for long-term evolution (LTE) handsets
[50]	$50 \times 100\,mm^2$	803–823 MHz 2440–2900 MHz	−17 dB @ lower band −9 dB @ higher band	Four-shaped monopole	Spiral loaded dumbbell DGS	Dual	<0.21	−4 dBi (Lower band) 2.4 dBi (Higher band)	MIMO for LTE application

(Continued)

Table 7.2 (Continued)

References	Dimension	Operational band	Port-to-port isolation $(S_{ij}, i \neq j)$	Antenna type	Type of DGS	No. of ports	ECC	Gain/Efficiency	Applications
[51]	$22 \times 36\,mm^2$	3.1–11 GHz with notch at 5.15–5.85 GHz	Less than −15 dB	Square monopole antenna	Slot in T-shaped ground	Dual	<0.1	1–5 dBi over the band	Portable UWB devices
[52]	$27 \times 30\,mm^2$	3–11 GHz with notch at 3.30–3.70 GHz and 5.15–5.85 GHz	Less than −20 dB	U-shaped radiator with metal strip	Vertical slot	Dual	<0.012	Efficiency >65%	UWB MIMO application
[53]	$32 \times 32\,mm^2$	3.1–10.6 GHz	Less than −15 dB	Open L-shaped slot	Single narrow slot	Dual	<0.04	1.7–4 dBi Efficiency > 60%	UWB MIMO application
[54]	$60 \times 60\,mm^2$	Around 2.45 GHz	Less than −25 dB	Square patch and ring antenna	Square ring defect	Four	<0.1	—	MIMO antenna
[55]	$70 \times 90\,mm^2$	2.45 and 5.2 GHz WLAN band	−33 dB	Monopole	Slotted CSRR	Dual	<0.016	4.025 dBi/ 86.6%	WLAN
[56]	$22 \times 26\,mm^2$	3.1–10.6 GHz	Less than −18 dB	Stepped slot antenna	T-shaped slot	Dual	<0.004	3.8 dBi/ 90%	UWB MIMO application
[57]	$30 \times 30\,mm^2$	2.28–7 GHz	Less than −14 dB	Triangular patch with L-shaped stub	T-shaped and rectangular slot	Dual	—	4–5 dBi	WLAN/ Wi-Fi/LTE

Ref.	Size	Frequency/Bandwidth	Isolation	Structure	Decoupling	Ports	Correlation	Gain/Efficiency	Application
[58]	60×60 mm²	2.45 GHz, 75 MHz BW	−22 dB	Concentric square ring patches	CSRR	Four	<0.4	Inner ring 4 dBi @ 84.6%, Outer ring 3.4 dBi, @ 72.5%	MIMO antenna at ISM band
[59]	60×130 mm²	2.45/5.2/5.8 GHz	Less than −15 dB	L-shaped radiator with interdigital shorting stubs	L and spiral slots	Four	<0.25	1.3–4.5 dBi, Efficiency: 40% @2.45 GHz 70–80% @ 5.2 GHz and 5.8 GHz	Mobile terminal
[60]	50×90 mm²	1.85–10.6 GHz	Less than −10 dB	Semi ring antenna	Slot and fork shaped	Dual	<0.16	1.22 dBi/51.8%	GSM/LTE/ WLAN
[61]	28×22 mm²	2.9–11.8 GHz	Less than −20 dB	Semi elliptical slot	H-shaped slot	Dual	<0.03	—	UWB MIMO application
[62]	33×26 mm²	3–11 GHz with notch at 4.7–5.5 GHz	Less than −15 dB	Semi-circular monopole slots	T-shaped open slot and rectangular slot	Dual	<0.03	1–6 dBi	UWB MIMO application
[63]	55×90 mm²	2.45 GHz band	−20 dB @2.4 GHz	PIFA	T slot with capacitor	Dual	–	2.1 dBi/40%	MIMO WLAN
[64]	23×29 mm²	3–12 GHz	Less than −15 dB	Triangular patch	CSRR and stubs	Dual	<0.15	1.2–5.9 dBi/ 82%	MIMO UWB
[67]	$\pi \times 80^2 \times 50$ mm³	LTE700/850/900 (698–960 MHz)	−10 to −15 dB	Monopole antenna with side sleeves	Meandered with slots	Four	<0.075	—	Small LTE base station

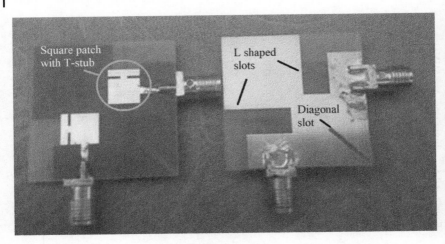

Figure 7.20 Photograph of the fabricated prototype showing two mutually perpendicular square patches with L-shaped slots along with diagonal defects on the ground plane. Source: Ren et al. [53] IEEE.

other. Simplicity in shape and use of slot DGSs are also quite interesting in ref. [53]. The DGS takes the shape of square loop in ref. [54] and slotted complementary split ring resonator (CSRR) in ref. [55]. The prototype examined in ref. [55] is shown in Figure 7.21 that demonstrates 33 dB element to element isolation along with miniaturization in the element size.

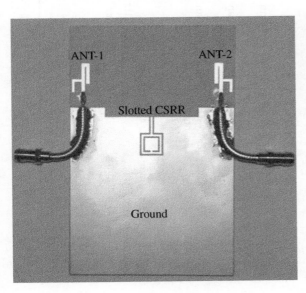

Figure 7.21 Photograph of the fabricated prototype showing two adjacent monopoles with slotted CSRR on the ground plane. Source: Dae-Geun Yang et al. [55] John Wiley & Sons, Inc.

Four port [59] and two port [60] MIMO antennas for mobile terminals exploited different DGS slots. A semi-elliptical slot antenna used H-shaped DGS [61] to realize a very compact UWB radiator. Another interesting design of two element PIFA operating around 2.45 GHz explored T-shaped defect in between them and placed an external capacitor to realize a resonant circuit [63]. This resulted in high isolation ~20 dB.

Base station MIMO antennas are not an exception. A 3D antenna structure was developed for LTE700/850/900 applications [67]. That explored intricate DGS geometries for a four-element base station antenna with elements to element spacing ~0.24λ [67].

The use of DRA in MIMO system is relatively uncommon, but the DRA researchers have been exploring the possibilities [65, 68–70]. A mm-wave design [68] employed two DRAs at 60 GHz that introduced a wideband FSS wall to isolate two DRA elements and resonant DGS slots to control the mutual coupling by surface currents. It obtained about 30 dB reduction over 57–63 GHz operating band. Dual band double port DRAs have also been used for worldwide interoperability for microwave access (WiMAX) (3.5 GHz) and WLAN (5.2 GHz) applications [69, 70] with a promise for 13–20 dB port isolation. A three-port two element DRA was reported in ref. [65], and its configuration is shown in Figure 7.22. The port configurations are self-explanatory. Two half ring DGS in between the elements provide about 20 dB isolation [65].

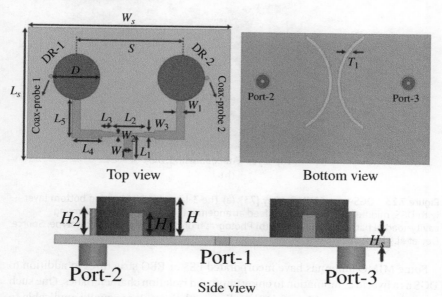

Figure 7.22 Schematic diagram of a triple port MIMO DRA integrated with DGS. Source: Das et al. [65] John Wiley & Sons.

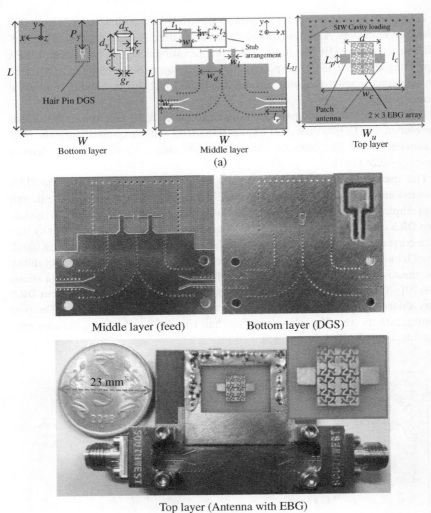

Figure 7.23 DGS-based MIMO array [73]: (a) The 3-layer layout showing bottom layer with DGS, middle layer bearing the feed arrangement, and the top layer carrying cavity-loaded patch and EBG array; (b) Photograph of the implemented prototype. Source: Dey et al. [73] IEEE.

Some MIMO antennas have incorporated FSS or EBG structures in addition to DGS as a hybrid combination to enjoy improved isolation characteristics. One such application is observed in ref. [68] as discussed above. It is equally applicable to the printed antennas [66, 71].

Like mm-wave MIMO DRA [68], printed antennas have also been explored in mm-wave domain [72–74]. A 5G antenna [73] operates from 27.5 to 28.35 GHz and the architecture is shown in Figure 7.23. Two square patches loaded with substrate integrated waveguide (SIW) cavity work as the radiator. The required isolation over the full band has been achieved by a hybrid technique. An EBG is visible in between two patches that takes care of wideband decoupling from 26.2 to 32.03 GHz. The order of isolation thus improves by about 13.9 dB. In addition, a hairpin DGS is visible at the bottom layer that results in an improvement of isolation by about 47.7 dB around 27.94 GHz. The peak isolation has been recorded as 71.9 dB with peak gain of 9 dBi [73]. The ECC is as low as 0.000 15 with diversity gain of 9.99.

A 5G mm-wave antenna-in-package has also used DGS [74] in realizing a 4 × 4 dual polarized patch array. A single unit of that multi-layered design [74] is shown in Figure 7.24. A prototype operates over 26.5–29.5 GHz with 17.37 dBi peak gain and more than 22.63 dB isolation over the frequencies of interest.

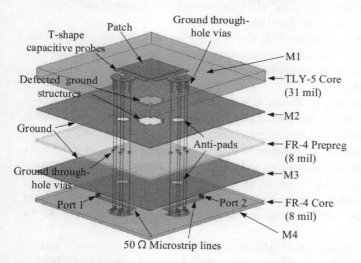

Figure 7.24 Layer wise layout of a single unit of a dual polarized DGS-based antenna reported. Source: Dey et al. [73] IEEE/CC BY 4.0.

7.6 Conclusion

The tremendous potential of DGS is revealed from the above applications and the related information. An antenna engineer nowadays can confidently remark that no 5G device today is possible to be realized for commercial applications without any DGS-based circuits and antennas. It is not only less expensive, but is extremely convenient in terms of its demand for no extra volume or space! The application areas are new and highly promising for the future. Thus one should expect more innovations in DGS-based compact antenna designs especially for 5G/6G systems along with new generation phased arrays.

References

1 M. A. Khayat, J. T. Williams, D. R. Jackson, and S. A. Long, "Mutual coupling between reduced surface-wave microstrip antennas," *IEEE Transactions on Antennas and Propagation*, vol. 48, no. 10, pp. 1581–1593, 2000.

2 R. P. Jedlicka and K. R. Carver, "Mutual coupling between microstrip antennas," *Proceedings of the Workshop on Printed Circuit Antenna Technology Held at Las Cruces, New Mexico*, pp. 4/1–19, October 1979.

3 R. P. Jedlicka, M. T. Poe, and K. R. Carver, "Measured mutual coupling between microstrip antennas," *IEEE Transactions on Antennas and Propagation*, vol. 29, no. 1, pp. 147–149, 1981.

4 D. R. Jackson, J. T. Williams, A. K. Bhattacharyya, R. L. Smith, S. J. Buchheit, and S. A. Long, "Microstrip patch antenna designs that do not excite surface waves," *IEEE Transactions on Antennas and Propagation*, vol. 41, no. 8, pp. 1026–1037, 1993.

5 D. Guha, S. Biswas, M. Biswas, J. Y. Siddiqui, and Y. M. M. Antar, "Concentric ring-shaped defected ground structures for microstrip applications," *IEEE Antennas and Wireless Propagation Letters*, vol. 5, pp. 402–405, 2006.

6 D. M. Pozar and D. H. Schaubert, "Analysis of an infinite array of rectangular microstrip patch with idealized probe feed," *IEEE Transactions on Antennas and Propagation*, vol. 32, no. 10, pp. 1101–1107, 1984.

7 D. M. Pozar and D. H. Schaubert, "Scan blindness in infinite phased arrays of printed dipoles," *IEEE Transactions on Antennas and Propagation*, vol. 32, no. 6, pp. 602–610, 1984.

8 A. H. Mohammadian, N. M. Martin, and D. W. Griffin, "A theoretical and experimental study of mutual coupling in microstrip antenna arrays," *IEEE Transactions on Antennas and Propagation*, vol. 37, no. 10, pp. 1217–1223, 1989.

9 J. R. James and A. Henderson, "High-frequency behavior of microstrip open-circuit terminations", *Microwaves, Optics and Acoustics*, vol. 3, no. 5, pp. 205–218, 1979.

10 E. H. V. Lil and A. R. V. de-Capelle, "Transmission line model for mutual coupling between microstrip antenna," *IEEE Transactions on Antennas and Propagation*, vol. 32, no. 8, pp. 816–821, 1984.

11 M. D. Deshpande and M. C. Bailey, "Analysis of finite phased arrays of circular microstrip patches," *IEEE Transactions on Antennas and Propagation*, vol. 37, no. 11, pp. 1355–1360, 1989.

12 D. H. Schabuert and K. S. Yngvesson, "Experimental study of a microstrip array on high permittivity substrate," *IEEE Transactions on Antennas and Propagation*, vol. 34, no. 1, pp. 92–97, 1986.

13 G. P. Gauthier, A. Courtay, and G. H. Rebeiz, "Microstrip antennas on synthesized low dielectric-constant substrate," *IEEE Transactions on Antennas and Propagation*, vol. 45, no. 8, pp. 1310–1314, 1997.

14 I. Papapolymerou, R. F. Frayton, and L. P. B. Katehi, "Micro machined patch antennas," *IEEE Transactions on Antennas and Propagation*, vol. 46, no. 2, pp. 275–283, 1998.

15 J. S. Colburn and Y. Rahmat-Samii, "Patch antennas on externally perforated high dielectric constant substrates," *IEEE Transactions on Antennas and Propagation*, vol. 47, no. 12, pp. 1785–1794, 1999.

16 F. Yang and Y. R. Samii, "Mutual coupling reduction in microstrip antennas using electromagnetic band-gap structure," *Antennas and Propagation Society International Symposium*, vol. 2, pp, 478–481, 8–13th July 2001, Boston, MA, USA.

17 F. Yang and Y. R. Samii, "Microstrip antennas integrated with electromagnetic band-gap (EBG) structures: a low mutual coupling design for array applications," *IEEE Transactions on Antennas and Propagation*, vol. 51, no. 10, pp. 2936–2946, 2003.

18 A. A. F. Neyestanak, F. Jolani, and M. Dadgarpour, "Mutual coupling reduction between two microstrip patch antennas," *Proceedings of Canadian Conference on Electrical and Computer Engineering*, Niagara Falls, ON, Canada, pp. 739–742, 2008.

19 F. Jolani, A. M. Dadgarpour, and G. Dadashzadeh, "Reduction of mutual coupling between dual-element antennas with new PBG techniques," *Proceedings of 13th International Symposium on Antenna Technology and Applied Electromagnetics and the Canadian Radio Science Meeting, IEEE*, Banff, AB, Canada, pp. 1–4, 2009.

20 M. M. Bait-Suwailam, O. F. Siddiqui, and O. M. Ramahi, "Mutual coupling reduction between microstrip patch antennas using slotted-complementary split-ring resonators," *IEEE Antennas and Wireless Propagation Letters*, vol. 9, pp. 876–878, 2010.

21 D. Guha, S. Biswas, and C. Kumar, "Annular ring shaped DGS to reduce mutual coupling between two microstrip patches," *Applied Electromagnetics Conference AEMC 2009*, pp. 1–4, 2010, Kolkata, India.

22 M. Salehi and A. Tavakoli, "A novel low mutual coupling microstrip antenna array design using defected ground structure," *International Journal of Electronics and Communications,* vol. 60, no. 10, pp. 718–723, 2006.

23 D. Guha and Y. M. M. Antar, Eds., "Chapter 12 in Microstrip and Printed Antennas," *Wiley,* UK, 2011.

24 C. Y. Chiu, C. H. Cheng, R. D. Murch, and C. R. Rowell, "Reduction of mutual coupling between closely-packed antenna elements," *IEEE Transactions on Antennas and Propagation,* vol. 55, no. 6, pp. 1732–1738, 2007.

25 F.-G. Zhu, J.-D. Xu, and Q. Xu, "Reduction of mutual coupling between closely-packed antenna elements using defected ground structure," *Electronics Letters,* vol. 45, no. 12, pp. 601–602, 2009.

26 D. Guha, S. Biswas, T. Joseph, and M. T. Sebastian, "Defected ground structure to reduce mutual coupling between cylindrical dielectric resonator antennas," *Electronics Letters,* vol. 44, no. 14, pp. 836–837, 2008.

27 S. Biswas and D. Guha, "Isolated open-ring defected ground structure to reduce mutual coupling between circular microstrips: characterization and experimental verification," *Progress In Electromagnetics Research M,* vol. 29, pp. 109–119, 2013.

28 S. Biswas and D. Guha, "Stop-band characterization of an isolated DGS for reducing mutual coupling between adjacent antenna elements and experimental verification for dielectric resonator antenna array," *International Journal of Electronics and Communications (AEÜ)* vol. 65(4), pp. 319–322, 2013.

29 K. Wei, J. Li, L. Wang, Z. Xing, and R. Xu, "S-shaped periodic defected ground structures to reduce microstrip antenna array mutual coupling," *Electronics Letters,* vol. 52, no. 15, pp. 1288–1290, 2016.

30 K. Wei, J. Li, L. Wang, Z. Xing, and R. Xu, "Mutual coupling reduction by novel fractal defected ground structure bandgap filter," *IEEE Transactions on Antennas and Propagation,* vol. 64, no. 10, pp. 4328–4335, 2016.

31 Q. Li, M. Abdullah, and X. Chen, "Defected ground structure loaded with meandered lines for decoupling of dual-band antenna," *Journal of Electromagnetic Waves and Applications,* vol. 33, no. 13, pp. 1764–1775, 2019.

32 B. Qian, X. Chen, and A. A. Kishk, "Decoupling of microstrip antennas with defected ground structure using the common/differential mode theory," *IEEE Antennas and Wireless Propagation Letters,* vol. 20, no. 5, pp. 828–832, 2021.

33 B. Qian, X. Huang, X. Chen, M. Abdullah, L. Zhao, and A. A. Kishk, "Surrogate-assisted defected ground structure design for reducing mutual coupling in 2 × 2 microstrip antenna array," *IEEE Antennas and Wireless Propagation Letters,* vol. 21, no. 2, pp. 351–355, 2022.

34 D. Gao, Z.-X. Cao, S.-D. Fu, X. Quan, and P. Chen, "A novel slot-array defected ground structure for decoupling microstrip antenna array," *IEEE Transactions on Antennas and Propagation,* vol. 68, no. 10, pp. 7027-7038, 2020.

35 Y.-F. Cheng, X. Ding, W. Shao, and B.-Z. Wang, "Reduction of mutual coupling between patch antennas using a polarization-conversion isolator," *IEEE Antennas and Wireless Propagation Letters*, vol. 16, pp. 1257–1260, 2017, https://doi .org/10.1109/LAWP.2016.2631621.

36 H. Moghadas, A. Tavakoli, and M. Salehi, "Elimination of scan blindness in microstrip scanning array antennas using defected ground structure," *International Journal of Electronics and Communications (AEÜ)*, vol. 62, pp. 155–158, 2008.

37 D. B. Hou, S. Xiao, B. Z. Wang, L. Jiang, J. Wang, and W. Hong, "Elimination of scan blindness with compact defected ground structures in microstrip phased array," *IET Microwaves, Antennas and Propagation*, vol. 3, no. 2, pp. 269–275, 2009.

38 S. Xiao, M.-C. Tang, Y.-Y. Bai, S. Gao, and B.-Z. Wang, "Mutual coupling suppression in microstrip array using defected ground structure," *IET Microwaves, Antennas and Propagation*, vol. 5, no. 12, pp. 1488–1494, 2011.

39 C. K. Ghosh, S. Biswas, and D. Mandal, "Study of scan blindness of microstrip array by using dumbbell-shaped split-ring DGS," *Progress In Electromagnetics Research M*, vol. 39, pp. 123–129, 2014.

40 W. Croswell, T. Durham, M. Jones, D. Schaubert, P. Friederich, and J. Maloney, "Wideband antenna arrays," in C. A. Balanis, Ed, Modern Antenna Handbook, *Wiley*, Hoboken, NJ, 2008.

41 S. Xiao, C. Zheng, M. Li, J. Xiong, and B.-Z. Wang, "Varactor-loaded pattern reconfigurable array for wide-angle scanning with low gain fluctuation," *IEEE Transactions on Antennas and Propagation*, vol. 63, no. 5, pp. 2364–2369, 2015.

42 R.-L. Xia, S.-W. Qu, P.-F. Li, D.-Q. Yang, S. Yang, and Z.-P. Nie, "Wide-angle scanning phased array using an efficient decoupling network," *IEEE Transactions on Antennas and Propagation*, vol. 63, no. 11, pp. 5161–5165, 2015.

43 L. Gu, Y.-W. Zhao, Q.-M. Cai, Z.-P. Zhang, B.-H. Xu, and Z.-P. Nie, "Scanning enhanced low-profile broadband phased array with radiator-sharing approach and defected ground structures," *IEEE Transactions on Antennas and Propagation*, vol. 65, no. 11, pp. 5846–5854, 2017.

44 C. G. Kakoyiannis and P. Constantinou, "Reducing coupling in compact arrays for WSN nodes via pre-fractal defected ground structures," *2009 European Microwave Conference (EuMC)*, Rome, Italy, pp. 846–849, 2009.

45 Y. Jiang, Y. Yu , M. Yuan, and L. Wu, "A compact printed monopole array with defected ground structure to reduce the mutual coupling," *Journal of Electromagnetic Waves and Applications*, vol. 25, no. 14, 15, pp. 1963–1974, 2011.

46 S. Cui, Y. Liu, W. Jiang, and S. X. Gong, "Compact dual-band monopole antennas with high port isolation," *Electronics Letters*, vol. 47, no. 10, pp. 570–580, 2011.

47 S. Blanch, J. Romeu, and I. Corbella, "Exact representation of antenna system diversity performance from input parameter description," *Electronics Letters*, vol. 39, no. 9, pp. 705–707, 2003.

48 H. Li, J. Xiong, and S. He, "A compact planar MIMO antenna system of four elements with similar radiation characteristics and isolation structure," *IEEE Antennas and Wireless Propagation Letters*, vol. 8, pp. 1107–1110, 2009.

49 M. S. Sharawi, S. S. Iqbal, and Y. S. Faouri, "An 800 MHz 2 × 1 compact MIMO antenna system for LTE handsets," *IEEE Transactions on Antennas and Propagation*, vol. 59, no. 8, pp. 3128–3131, 2011.

50 M. S. Sharawi, A. B. Numan, M. U. Khan, and D. N. Aloi, "A dual-element dual-band MIMO antenna system with enhanced isolation for mobile terminals," *IEEE Antennas and Wireless Propagation Letters*, vol. 11, pp. 1006–1009, 2012.

51 L. Liu, S. W. Cheung, and T. I. Yuk, "Compact MIMO antenna for portable UWB applications with band-notched characteristic," *IEEE Transactions on Antennas and Propagation*, vol. 63, no. 5, pp. 1917–1924, 2015.

52 J. Li, Q. Chu, Z. Li, and X. Xia, "Compact dual band-notched UWB MIMO antenna with high isolation," *IEEE Transactions on Antennas and Propagation*, vol. 61, no. 9, pp. 4759–4766, 2013.

53 J. Ren, W. Hu, Y. Yin, and R. Fan, "Compact printed MIMO antenna for UWB applications," *IEEE Antennas and Wireless Propagation Letters*, vol. 13, pp. 1517–1520, 2014.

54 R. Anitha, V. P. Sarin, P. Mohanan, and K. Vasudevan, "Enhanced isolation with defected ground structure in MIMO antenna," *Electronics Letters*, vol. 50, no. 24, pp. 1784–1786, 2014.

55 D.-G. Yang, D. O. Kim, and C.-Y. Kim, "Design of dual-band MIMO monopole antenna with high isolation using slotted CSRR for WLAN," *Microwave and Optical Technology Letters*, vol. 56, no. 10, pp. 2252–2257, 2014.

56 C. Luo, J. Hong, and L. Zhong, "Isolation enhancement of a very compact UWB-MIMO slot antenna with two defected ground structures," *IEEE Antennas and Wireless Propagation Letters*, vol. 14, pp. 1766–1769, 2015.

57 W. N. N. W. Marzudi, Z. Z. Abidin, S. H. Dahlan, Ma Yue, R. A. Abd-Alhameed, and M. B. Child, "A compact orthogonal wideband printed MIMO antenna for WiFi/WLAN/LTE applications," *Microwave and Optical Technology Letters*, vol. 57, no. 7, pp. 1733–1738, 2015.

58 A. Ramachandran, S. Valiyaveettil Pushpakaran, M. Pezholil, and V. Kesavath, "A four-port MIMO antenna using concentric square-ring patches loaded with CSRR for high isolation," *IEEE Antennas and Wireless Propagation Letters*, vol. 15, pp. 15196–1199, 2016.

59 H. Huang, Y. Liu, and S.-X. Gong, "Four antenna MIMO system with compact radiator for mobile terminals," *Microwave and Optical Technology Letters*, vol. 57, no. 6, pp. 1281–1286, 2015.

60 S. K. Dhar and M. S. Sharawi, "A UWB semi-ring MIMO antenna with isolation enhancement," *Microwave and Optical Technology Letters*, vol. 57, no. 8, pp. 1941–1946, 2015.

61 L. Kang, H. Li, X.-H. Wang, and X.-W. Shi, "Miniaturized band-notched UWB MIMO antenna with high isolation," *Microwave and Optical Technology Letters*, vol. 58, no. 4, pp. 878–881, 2016.

62 G. Liu, Y. Liu, and S. Gong, "Compact uniplanar UWB MIMO antenna with band-notched characteristic," *Microwave and Optical Technology Letters*, vol. 59, no. 9, pp. 2207–2212, 2017.

63 J. Park, J. Choi, J.-Y. Park, and Y.-S. Kim, "Study of a T-shaped slot with a capacitor for high isolation between MIMO antennas," *IEEE Antennas and Wireless Propagation Letters*, vol. 11, pp. 1541–1544, 2012.

64 M. S. Khan, A.-D. Capobianco, S. M. Asif, D. E. Anagnostou, R. M. Shubair, and B. D. Braaten, "A compact CSRR-enabled UWB diversity antenna," *IEEE Antennas and Wireless Propagation Letters*, vol. 16, pp. 808–812, 2017.

65 G. Das, A. Sharma, R. K. Gangwar, and M. S. Sharawi, "Triple-port, two-mode based two element cylindrical dielectric resonator antenna for MIMO applications," *Microwave and Optical Technology Letters*, vol. 60, no. 6, pp. 1566–1573, 2018.

66 Z. Niu, H. Zhang, Q. Chen, and T. Zhong, "Isolation enhancement in closely coupled dual-band MIMO patch antennas," *IEEE Antennas and Wireless Propagation Letters*, vol. 18, no. 8, pp. 1686–1690, 2019.

67 Y.-S. Chen and C.-P. Chang, "Design of a four-element multiple-input–multiple-output antenna for compact long-term evolution small-cell base stations," *IET Microwaves, Antennas and Propagation*, vol. 10, no. 4 pp. 385–392, 2016.

68 R. Karimian, A. Kesavan, M. Nedil, and T. A. Denidni, "Low-mutual-coupling 60-GHz MIMO antenna system with frequency selective surface wall," *IEEE Antennas and Wireless Propagation Letters*, vol. 16, pp. 373–376, 2017.

69 A. A. Khan, M. H. Jamaluddin, S. Aqeel, J. Nasir, J. R. Kazim, and O. Owais, "A dual-band MIMO dielectric resonator antenna for WiMAX/WLAN applications," *IET Microwaves, Antennas and Propagation*, vol. 11, no. 1, pp. 113–120, 2017.

70 G. Das, A. Sharma, and R. K. Gangwar, "Dielectric resonator based two-element MIMO antenna system with dual band characteristics," *Antennas Propagation IET Microwaves*, vol. 12, no. 5, pp. 734–741, 2018.

71 Y. Liu, X. Yang, Y. Jia, and Y. J. Guo, "A low correlation and mutual coupling MIMO antenna," *IEEE Access*, vol. 7, pp. 127384–127392, 2019.

72 B. T. P. Madhav, Y. U. Devi, and T. Anilkumar, "Defected ground structured compact MIMO antenna with low mutual coupling for automotive communications," *Microwave and Optical Technology Letters*, vol. 61, no. 3, pp. 794–800, 2019.

73 S. Dey, S. Dey, and S. K. Koul, "Isolation improvement of MIMO antenna using novel EBG and hair-pin shaped DGS at 5G millimeter wave band," *IEEE Access*, vol. 9, pp. 162820–162834, 2021.

74 W. Kim, J. Bang, and J. Choi, "A cost-effective antenna-in-package design with a 4 × 4 dual-polarized high isolation patch array for 5G mm wave applications," *IEEE Access*, vol. 9, pp. 163882–163892, 2021.

8

DGS Applied to Circularly Polarized Antenna Design

8.1 Introduction

The general trend of investigations on DGS-based antennas has been popular in exploring linearly polarized (LP) printed antennas. But in reality, the designs of circularly polarized (CP) antennas are more challenging and widely used in several applications. The CP antennas are immune to multipath reflections which are of a great significance in personal communication systems [1]. They hardly get influenced by the rotation of the polarization vector, commonly known as Faraday rotation, in the intervening medium. As a result, they are very much indispensable in many areas like multipath environment, low earth orbit space communication, and global satellite navigation service. Several expensive techniques, in terms of cost, real estate requirement, and complexity are commonly employed to realize useful CP performance over a considerably wide bandwidth. The DGS-integration technique has also been successfully explored in CP antenna design to combat those challenges. Its low profile, low cost, and light weight features have made the technique more attractive for the space and air borne systems. Yet only a limited number of patents and research publications indicate that the real potential of DGS for CP antenna designs has not been adequately explored.

8.2 Basic Principle of CP Generation in a Microstrip Patch

The principle of superposition of two mutually orthogonal time varying field vectors is followed in designing a CP antenna. As illustrated in Figure 8.1, $\mathbf{R}$ is the resultant vector of a pair of two such vectors $\mathbf{A_x}$ and $\mathbf{B_y}$ expressed as

Defected Ground Structure (DGS) Based Antennas: Design Physics, Engineering, and Applications,
First Edition. Debatosh Guha, Chandrakanta Kumar, and Sujoy Biswas.
© 2023 The Institute of Electrical and Electronics Engineers, Inc. Published 2023 by John Wiley & Sons, Inc.

$|\mathbf{A_x}| = a_x \, sin(\omega t + \varphi)$ and $|\mathbf{B_y}| = b_y \, sin(\omega t)$ with mutual phase difference φ. Then $|\mathbf{R}|$ is given by

$$|\mathbf{R}| = \sqrt{a_x^2 sin^2(\omega t + \varphi) + b_y^2 sin^2 \omega t} \qquad \text{for } \theta = 90° \qquad (8.1)$$

and the angle Ψ is given by

$$\Psi = \tan^{-1}\frac{b_y sin\omega t}{a_x sin(\omega t + \varphi)} \qquad (8.2)$$

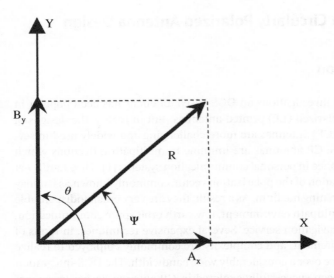

Figure 8.1 A scheme of CP antenna design expressed in terms of the rotating vector R as a function of time originated from the contributing vectors $\mathbf{A_x}$ and $\mathbf{B_y}$ under a set of given conditions.

Equation (8.2) thus represents Ψ as a function of time with $\mathbf{R}$ rotating about the axis perpendicular to the plane bearing $\mathbf{A_x}$ and $\mathbf{B_y}$. Orthogonality of the vectors ($\theta = 90°$) with $a_x = b_y$ and $\varphi = 90°$ (LHCP: left hand circularly polarized) or $-90°$ (RHCP: right hand circularly polarized) are the ideal conditions required for an ideal CP operation. Corresponding axial ratio ($AR = |\mathbf{A_x}|/|\mathbf{B_y}|) = 1$. Any deviation from these conditions leads to an elliptical polarization with $AR > 1$.

This scheme is the basis of microstrip-based CP antenna design. The idea is to excite two orthogonal field vectors with the condition $a_x \approx b_y$ and mutual phase difference $\varphi \approx 90°$ as the ideal conditions are very difficult to achieve in practice. This introduces some degree of complexity in the antenna and feed structure. As discussed above, a deviation from the ideal conditions makes the resultant vector trace an "ellipse" instead of a "circle." This eventually results in an elliptically polarized radiation which is the most commonly occurring feature in a practical CP antenna.

In a single-element CP design, these two vectors $\mathbf{A_x}$ and $\mathbf{B_y}$ are realized in the form of two orthogonally polarized modes bearing electric fields $\mathbf{E_x}$ and $\mathbf{E_y}$, respectively. Typically, two mechanisms are followed for the same. In one method, a single traditional feed is used either with its strategic deployment or with strategic perturbation in the patch geometry as schematically shown in Figure 8.2 [2]. Each of the designs takes care of achieving $a_x \approx b_y$ with both θ and $\varphi \approx 90°$. The other way is to use two individual feeds to obtain $\theta = 90°$ along with a feed network to obtain $\varphi = 90°$ as shown in Figure 8.3 [3].

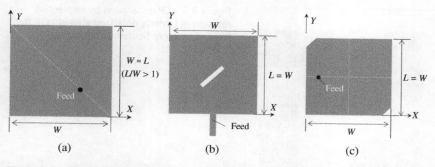

Figure 8.2 Single feed CP microstrip patches: (a) Near square patch fed by a coaxial probe on the diagonal plane; (b) edge-fed square patch loaded by a strategic slot; (c) truncated corner square patch fed by a coaxial probe on a principal axis. Source: Adapted from Balanis [2].

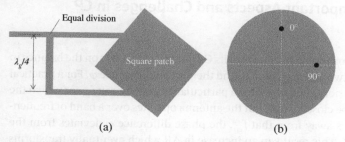

Figure 8.3 Producing CP radiation using dual feed with 90° phase difference: (a) square patch with microstrip line feed; (b) circular patch with coaxial probe feed. Source: Adapted from [2, 5].

Multiple feeds in a single element [4] and combinations of multiple elements [4–6] in the form of array are also used to generate more than two E-field vectors and their befitting relative phases. The examples are furnished through Figures 8.4 and 8.5 and they are self-explanatory. Higher order mode-based CP designs are also possible [7].

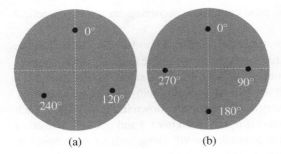

Figure 8.4 CP design of a circular microstrip patch by multipoint feeding with appropriated phase relation: (a) use of 3 probes; (b) use of 4-probes Source: Adapted from [2, 5].

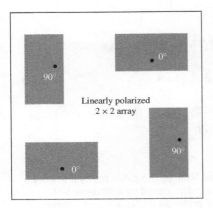

Figure 8.5 CP generation by sequential rotation of linearly polarized elements in the form of an array. Source: Adapted from Huang [5].

8.3 Some Important Aspects and Challenges in CP Designs

It is already mentioned that the purity of its CP radiation depends on the balance in the magnitude of two vectors a_x and b_y and their phase difference φ. For a practical antenna, this criterion could be met at a particular frequency, say f_{cp} for which the axial ratio would be close to 0 dB. But the antenna operates over a band of frequencies. As one moves away from that f_{cp}, the phase difference φ deviates from the target value of 90°. This results in an increase in AR which eventually transforms the circular polarization to elliptical polarization. This deviation is acceptable to a certain extent and is typically limited by 0 dB < AR < 3 dB. For high precision dual CP links, AR below 1 dB is recommended whereas for a wide beam or handheld system, up to 10 dB or more may be acceptable. A practicing engineer, therefore, focuses on the range of frequencies over which the AR remains below the specified value and that range of frequencies determines its axial ratio bandwidth (AR bandwidth). Typically, AR varies too fast with frequency resulting in narrow AR bandwidth and that is one of the major challenges of a CP antenna design.

But primarily, an antenna has to be matched with the feed resulting $S_{11} < -10\,\text{dB}$ over a specified frequency range, commonly termed as *impedance bandwidth* or *matching bandwidth*. Thus, the common frequencies satisfying both *impedance matching* as well as the required AR *value* determine the *useable operating band* for a CP antenna. In turn, it imposes a critical condition that *the matching bandwidth needs to accommodate the AR bandwidth*. But in practice, even after executing all feasible technical exercises it is found that the best possible axial ratio and the best achievable impedance matching may occur at two different frequency ranges that are widely separated from each other. This leads to the worst situation when no mutual overlap exists and hence the design offers effectively "zero" CP bandwidth [8]. This is truly a major challenge and hence *wideband* CP antenna is all-time challenge to a design engineer. Another aspect is to maintain the antenna patterns as symmetric as possible since the broad banding techniques sometimes render the radiation patterns disturbed [9]. Wide CP bandwidth is achievable by different techniques and one of them [10] claimed as high as 82% AR bandwidth (over which AR remains below 3 dB). That work [10] actually used a complicated four-point "*L*"- type proximity feeding supported by Wilkinson power divider and switched-network. This merit of large bandwidth is associated with multiple limitations like feed complexities, increased profile and footprint, non-suitability for polytetrefluoroethylene (PTFE) substrate-based design, and higher cost. In the present context, application of DGS will be considered in facilitating improved CP antenna designs.

8.4 DGS Integrated Single-Fed CP Antenna Design

In a single fed patch, the specific requirement of orthogonality in resonant fields or currents is typically realized by perturbing the patch geometry [8, 9, 11]. Based on adequate knowledge and experience gathered since 2005 [12], the antenna researchers started applying DGS to CP designs without any tampering or perturbation on the patch surface. Variety of techniques bearing several versatilities has been discussed here.

8.4.1 Use of Slot-Type DGS

Defect embedded ground plane for microstrip CP antenna designs began as early as 2007 [13]. This was before a DGS-based design became popular as a standard practice. The antenna geometry used in [13] is shown in Figure 8.6. The ground plane bears cross-slots of unequal length with a strategically located probe. This cross-slot DGS acts as a perturbing agent to produce a pair of orthogonal fields required for CP generation. The slot dimensions determine the sense of

CP as RHCP with $L_1 > L_2$ and LHCP with $L_1 < L_2$, the feed location remaining unchanged. The purpose of using annular rings is to obtain dual resonances as shown in Figure 8.7a. The matching bandwidth at each frequency is about 6% although its AR bandwidth turns out to be nearly 1% around each resonance. A representative radiation pattern at 1.224 GHz is shown in Figure 8.7b with 1.35 dBic peak gain. The CP operation over an angular span of about 50° is revealed.

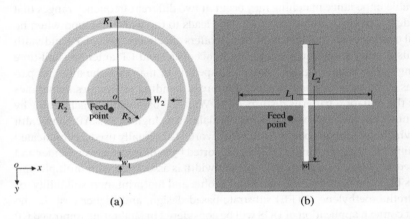

(a)   (b)

Figure 8.6 Dual frequency annular ring CP antenna with DGS integrated ground plane [13]: (a) Top view; (b) back side view. Source: Bao and Ammann [13] © IEEE.

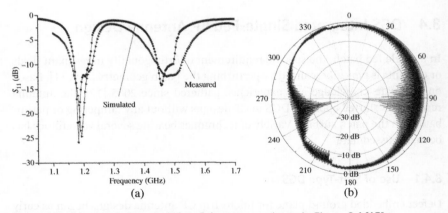

(a)   (b)

Figure 8.7 Measured characteristics of the antenna shown in Figure 8.6 [13]: (a) Reflection coefficient versus frequency; (b) radiation pattern obtained at 1.224 GHz with spinning dipole as the radiator. Parameters: $R_1 = 24$ mm, $R_2 = 18.1$ mm, $R_3 = 6.5$ mm, $W_1 = 0.8$ mm, $W_2 = 6.3$ mm, $w = 1$, $L_1 = 40$ mm, $L_2 = 42.4$ mm, feed at 3 mm away from the center. Substrate thickness 1.52 mm with $\varepsilon_r = 4$, ground plane size 60 mm × 60 mm [13]. Source: Bao and Ammann [13] © IEEE.

8.4.2 Use of Fractal DGS

Fractal-shaped DGS has been found to be useful in CP design [14]. Figure 8.8 provides a schematic view of the same. The progressive order of the fractal geometry is shown in Figure 8.8a. A second or the third order fractal DGS has been deployed adjacent to the patch edge and closer to the probe maintaining a distance d. This indeed is a radiating edge and apparently, the patch configuration looks a linearly polarized antenna. Thus, the fringing fields around this edge should be predominantly strong due to the proximity of the feeding probe and hence the ground plane

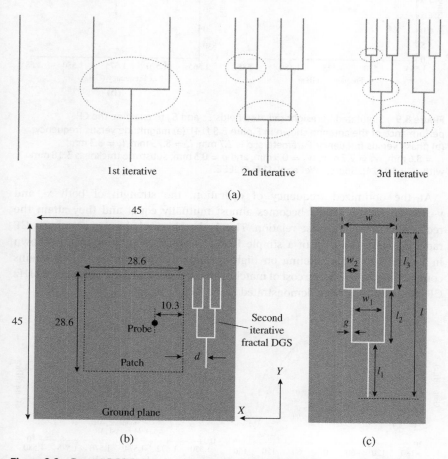

1st iterative   2nd iterative   3rd iterative

(a)

(b)   (c)

Figure 8.8 Fractal DGS for generating CP operation in a square microstrip patch [14]: (a) fractal geometry from 0th order to 3rd order; (b) back side view of a DGS integrated square patch for RHCP radiation; (c) enlarged view of the 2nd order fractal DGS [14]. Source: Wei et al. [14] © IEEE.

current. The purpose of the DGS is to create orthogonal current components on the ground plane. The square patch naturally supports the orthogonal resonance fields and enhances the strength. This insight has been ensured through a study depicted in Figure 8.9.

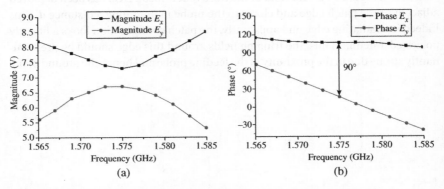

Figure 8.9 Simulated boresight radiated fields E_x and E_y to examine the CP performance of the antenna shown in Figure 8.8 [14]: (a) magnitude versus frequency; (b) phase versus frequency. Parameters: $d = 2.7$ mm, $l_1 = 8.75$ mm, $l_2 = 8.3$ mm, $l_3 = 8.6$ mm, $w_1 = 2.2$ mm, $w_2 = 0.9$ mm, and $g = 0.3$ mm; substrate thickness 3.18 mm with $\varepsilon_r = 10$ [14]. Source: Wei et al. [14] © IEEE.

At the optimized frequency of operation, the strength of both x- and y-components of E field becomes almost mutually equal and they attain the required orthogonal phase relation. This is the physics behind generating a CP radiation with the help of a simple DGS. Its radiation performance is shown in Figure 8.10. This antenna on high permittivity substrate ($\varepsilon_r = 10$) attains compactness in size at the cost of matching bandwidth and gain. However, 6 MHz CP bandwidth has been demonstrated at the GPS L_1 band.

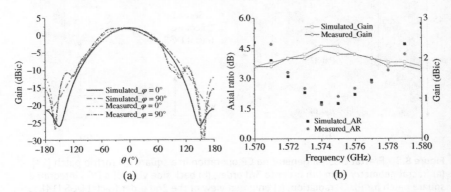

Figure 8.10 Characteristics of the antenna shown in Figure 8.8 [14]: (a) co-polarized radiation patterns at 1575 MHz; (b) Peak gain and axial ratio versus frequency. Parameters as in Figure 8.9 [14]. Source: Wei et al. [14] © IEEE.

This fractal DGS has been used in a more versatile way to generate LHCP/RHCP pair using a scheme as shown in Figure 8.11 [15]. Here, a single DGS serves a pair of patches and they are subjected to mutually orthogonal sense of surface current distribution. That results in a polarization diversity CP antenna pair and should be useful in operating two communication links simultaneously with the same frequency.

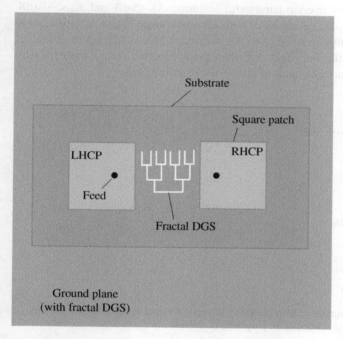

Figure 8.11 Polarization diversity CP antenna based on a single unit of fractal DGS. Source: Adapted from Wei et al. [15].

8.4.3 Use of Grid DGS

The concept of grid DGS has been introduced in Chapter 4. That discussion had focused on its application to reducing XP radiations. The final shape of the grid, as explained earlier, is obtained by configuring the interconnects in between the grid elements (Figure 4.35). For a CP design, the grid DGS is configured to support the orthogonal fields to grow and maintain a phase relation of about 90° with respect to the primary resonant fields. This requirement in principle is just opposite to that aimed in suppressing XP radiations.

A CP square patch using a grid DGS is shown in Figure 8.12 [16]. The orthogonal degenerate modes obtained in a square-shaped patch are of typical interest for CP generation. The grid configuration in Figure 8.12 is a variant of that shown in Figure 4.35c and facilitates LHCP radiations. The grid pair in Figure 4.35c was employed to reduce XP fields and followed mirror image pattern. That configuration is somewhat different from the preset one used in Figure 8.12b. The original geometry [16] leaves all 62 interconnects available for optimization. The antenna was designed to operate around 3.55 GHz with about 2.2% AR bandwidth. Figure 8.13a gives an idea about its AR values and CP gain. Both measured and simulated data show a consistent peak gain of about 6.6 dBic covering the overlap bandwidth featuring AR <3 dB and $S_{11} < -10$ dB. The AR values are also functions of both elevation angle θ and radiation plane φ as examined in Figure 8.13b. The AR maintains below 3 dB up to about 50° elevations around the boresight and the lowest reported AR is about 0.08 dB which indeed appears as a mark of purity of CP performance.

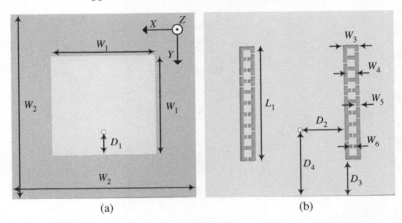

Figure 8.12 CP design using grid-DGS [16]: (a) top view; (b) viewed from the ground plane side [16]. Source: Zhang et al. [16] IEEE/CC BY-4.0.

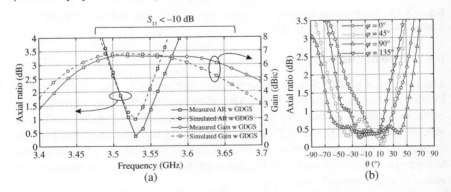

Figure 8.13 CP and radiation characteristics of the antenna shown in Figure 8.12 [16]: (a) on axis AR and gain versus frequency; (b) Axial ratio (AR) as a function of the elevation angle theta for different radiation planes [16]. Source: Zhang et al. [16] IEEE/CC BY-4.0.

8.4.4 Use of PIN Switch Integrated Reconfigurable DGS

As discussed, a defect on the ground plane means a perturbation to the surface current and associated fields. They could be electronically reconfigured using some active devices like p-i-n (PIN) diodes across the slots [17]. The diodes along with biasing networks can be safely housed behind the antenna structure without causing any disturbance to the radiated fields. A few such designs [18, 19] are discussed here.

One of them is shown in Figure 8.14 [18] which uses a pair of antipodal L-slits beneath a microstrip patch. They excite LHCP and the mirror image of the defects

Figure 8.14 PIN switch integrated DGS for CP radiation [18]: (a) a square patch with "L" - shaped slit DGS producing LHCP radiations; (b) 4 DGS units integrated with switchable PIN diodes D_{1-8} [18]. Source: Yoon et al. [18] © IEEE.

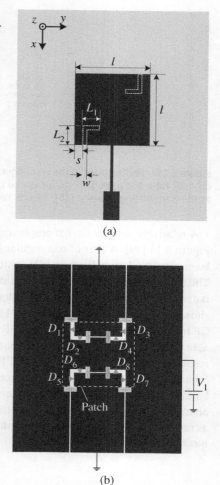

would produce RHCP in the same patch. A combination of four slits integrated with 8 PIN diodes D_{1-8} is shown in Figure 8.14b. These PIN diodes work as switches to turn the DGSs from open to short and vice versa as per the design protocol and demonstrate reconfigurable dual CP operation [18]. The antenna performance is depicted in Figure 8.15. Measured AR bandwidth for both LHCP and RHCP is about to be 30 MHz with about 3 dBic peak gain. The radiation pattern measured with a rotating dipole as the transmitting radiator also endorses the purity of circular polarization around the boresight.

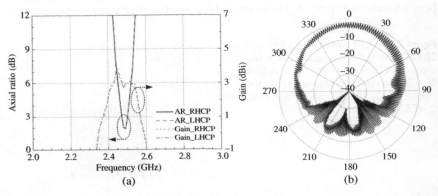

Figure 8.15 Performance of the reconfigurable CP antenna shown in Figure 8.14b [18]: (a) Axial ratio and peak gain; (b) measured LHCP radiation pattern using rotating dipole as the transmitting antenna [18]. Source: Yoon et al. [18] © IEEE.

A relatively simpler design employing only a pair of PIN switches is shown in Figure 8.16 [19]. A pair of rectangular loop DGSs near two strategic corners has been employed and they use PIN switches P1 and P2 to operate in a CP mode. The same element with no DGS should resonate with its fundamental mode and radiate linearly polarized fields. An identical phenomenon occurs in the antenna shown in Figure 8.16 too when both P1 and P2 are in ON state. The patch works in CP mode when either one goes to OFF state and its associated DGS causes favorable perturbation to the surface current. To be more specific, P1 OFF with P2 ON results in LHCP and P1 ON with P2 OFF generates RHCP [19]. This reconfigurable S-band design performs equally attractive as shown in Figure 8.17 [19]. The AR bandwidth is about 20 MHz for both senses of CP with the best achievable value of about 0.5 dB. The purity of the CP is also reflected in its radiation pattern showing as much as 6.4 dBic peak gain. The same spinning linear measurement technique has been used here.

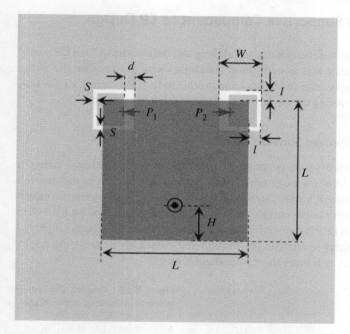

Figure 8.16 PIN Switch Integrated loop DGS beneath two strategic corners of a square patch for CP operation. Source: Adapted from Yang et al. [19].

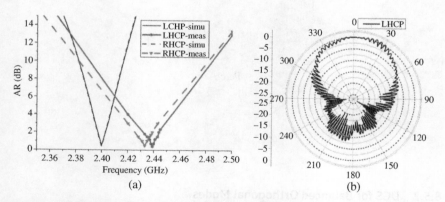

Figure 8.17 Performance of the reconfigurable CP antenna shown in Figure 8.16 [19]: (a) axial ratio versus frequency (b) measured LHCP radiation pattern at the mid-band frequency [19]. Source: Yang et al. [19] © IEEE.

8.5 DGS as a Supportive Component to CP Design

The DGS sometimes indirectly helps in fulfilling the requirements for a CP patch. Its reactive impedance acts as additional matching element to realize the phase and amplitude balance over the target bandwidth. On the other hand, a DGS may also help in maneuvering the surface currents favoring CP operation.

8.5.1 DGS for Improved Surface Current

A circularly polarized patch generates a rotating substrate field and hence a rotating surface current on the ground plane. A symmetry in current distribution actually controls the AR value. But structural asymmetry is a common concern that causes asymmetry in the surface currents. The asymmetric feed for a 2.3 GHz patch in Figure 8.18 [20] is one such example. Five dot DGSs have been employed to regain the symmetry in current distribution and eventually improve the AR by 1 dB. This effectively enhances the CP bandwidth by 25%.

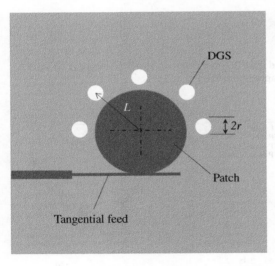

Figure 8.18 Tangentially fed CP antenna integrated with circular dot-shaped DGSs. Source: Adapted from Lee et al. [20].

8.5.2 DGS for Balanced Orthogonal Modes

A simple design described in Figure 8.6 is an ideal example of applying DGS to obtain a balanced pair of orthogonal modes in terms of equal amplitude and phase quadrature. This, as mentioned earlier, is the key requirement behind in a CP patch design. The DGS, in several such designs, has also been explored as a useful method in improving that feature. Figure 8.19 represents such an example [21] which bears a two layer composition. The top layer (Figure 8.19a) is a proximity coupled L-band circular patch loaded with a tilted cross-slot. The purpose of the cross-slot is to generate two mutually orthogonal resonances [21]. The bottom layer beneath the patch as well as the feed line is the common ground plane which bears an additional fractal DGS (Figures 8.19b,c). This DGS actually facilitates matching over a larger frequency band maintaining the required amplitude and phase relations among those orthogonal modes. This eventually enhances the AR bandwidth by about 62%, impedance bandwidth by 70%, and the radiation efficiency by 4% relative to its without DGS configuration. The capacitive loading in turn reduces the size of the antenna by 45% at the cost of gain reduction by 0.9 dB.

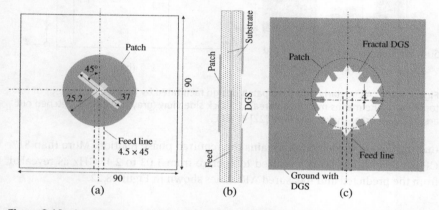

Figure 8.19 Improvement of CP performance by DGS integration technique [21]: (a) top view of a proximity fed circular patch; (b) cross sectional view showing two distinct layers; (c) bottom view of the common ground plane bearing a fractal-DGS. Source: Adapted from Prajapati et al. [21].

In some cases, dual feeding is used for generating two orthogonal modes. Figure 8.20a shows one such 2.18 GHz design [22] where a branch-line coupler is used as the feed network. A challenge in such printed circuits is to achieve the required line impedance. Use of "H"-shaped DGS beneath the lines resolve the issue as shown in this layout of Figure 8.20b [22]. This helps in maintaining the compact network within the limited available space. The DGS integration also improves the operating bandwidth by a factor of 2.

A CP design can also use more number of feeds to achieve large CP bandwidth but at the cost of complicated feed and power divider networks [10, 23, 24]. One such approach for circular microstrip patch is shown in Figure 8.21 [23]. It uses four L-shaped probes with a wideband balun comprising a 3-dB Wilkinson power divider and a broadband Schiffman phase shifter. A loop-shaped DGS is also a part of the design. It actually helps the even mode capacitance to decrease faster than

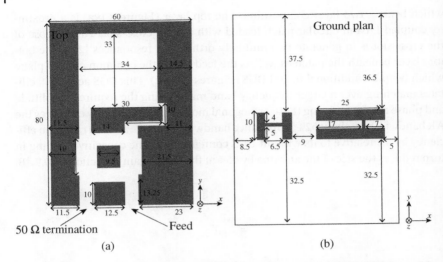

Figure 8.20 A circularly polarized patch design facilitated by DGS [22]: (a) layout of for top view (white means metallized area); (b) back side view (gray scale means etched out area) [22]. Source: Thakur and Park [22] © IEEE.

that in the odd mode and maintains the required phase relation. More than 81% CP bandwidth has been produced to operate from 1.03 to 2.45 GHz as revealed from the predicted and measured AR values shown in Figure 8.22.

8.5.3 DGS for Optimizing CP Bandwidth

Corner truncated square patch with a single coaxial feed [8] as shown in Figure 8.23 [25] is probably the simplest configuration which is capable of generating orthogonal resonances producing circular polarizations. Its truncation parameter t plays a major role in controlling the orthogonal resonances with marginal off in the resonance frequencies. This results in relatively wider matching bandwidth ($S_{11} < -10$ dB) compared to a traditional linearly polarized patch. This impedance bandwidth can further be increased by increasing the substrate thickness [9, 26]. But the acceptable AR bandwidth of the antenna shows considerable offset relative to the matched band. The nature of the offset is studied in Figure 8.24 [25] which appears really challenging. The degree of offset enhances with an increase in the substrate thickness. A set of interesting studies with five CP antennas as in Table 8.1 etched on 62 mil RT Duroid 5870 substrate is summarized in Figure 8.24b. They operate at five different frequencies around 2.5, 3, 4, 6, and 10 GHz thus revealing five different h/λ_g values. It can be noted that the lowest h/λ_g ($= 0.02$) shows about 20 MHz AR bandwidth indicating almost no offset.

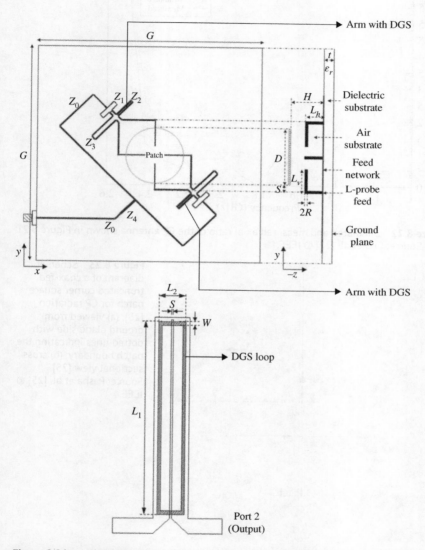

Figure 8.21 A circular patch fed by quadruple L-probe integrated with wideband balun comprising 3-dB Wilkinson power divider, broadband Schiffman phase shifter, and loop DGS [23] (enlarged view of the arm with DGS is shown in the bottom). Source: Guo et al. [23] © IEEE.

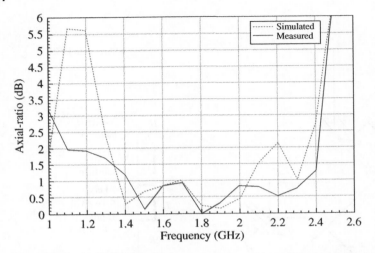

Figure 8.22 Simulated and measured axial ratio of the CP antenna shown in Figure 8.21 [23]. Source: Guo et al. [23] © IEEE.

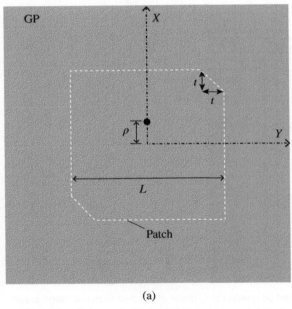

(a)

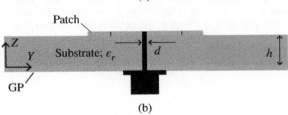

(b)

Figure 8.23 Schematic diagram of a coax-fed truncated corner square patch for CP radiation [25]: (a) viewed from ground plane side with dotted lines indicating the patch boundary; (b) cross sectional view [25]. Source: Pasha et al. [25] © IEEE.

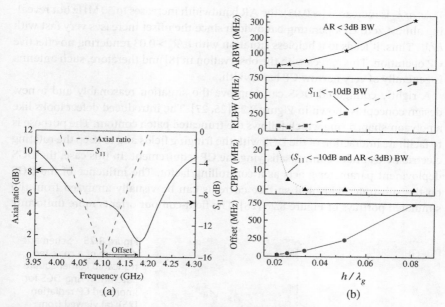

Figure 8.24 Characteristics of impedance (S_{11}) and axial ratio (AR) bandwidths of a truncated corner CP patch shown in Figure 8.23 [25]: (a) AR and the S_{11} responses indicating mutual offset over the frequencies of interest; (b) examination of AR and S_{11} bandwidths and the relative offset as a function of substrate thickness. Parameters as in (Table 8.1) [25]. RLBW = return loss bandwidth, CPBW = circular polarized bandwidth, ARBW = axial ratio bandwidth. Source: Pasha et al. [25] © IEEE.

Table 8.1 Design parameters for the antenna geometry in Figure 8.23 [25] (all dimensions are in mm).

Frequency of operation (GHz)	Patch size L (mm)	Probe location ρ (mm)	Patch truncation t (mm)	Other parameters
2.5	36.4	7.3	3.8	$d = 1.3$, $h = 1.575$,
3	29.6	4.4	3.5	$\varepsilon_r = 2.33$, ground
4	22.5	3.5	3	plane: $\lambda_0 \times \lambda_0$
6	14.22	2.5	2.5	
10	8.4	1.7	2	

Source: Pasha et al. [25] © IEEE.

As h/λ_g is increased (≈ 0.03), the AR bandwidth increases to 50 MHz but revealing almost zero CP operating bandwidth since the offset increases very fast with h/λ_g. Thus, it leads to a helpless situation with $h/\lambda_g > 0.03$ rendering no effective CP operation. This agrees with the observation in [8] and therefore, such antenna is generally of very narrow CP bandwidth.

A rightly configured DGS can improve the situation reasonably and a new design concept is shown in Figure 8.25 [25, 27]. The introduced defect looks like a polygon-shaped ring which follows the truncated patch contour. The purpose is to facilitate interaction of the DGS with the fringing fields and to keep the rotating electric fields under control satisfying the CP requirement. In this case, the DGS deployment parameter g acts as a controlling factor. The influence of the DGS on the substrate fields and surface currents can be visually analyzed from the simulated portrays of Figure 8.26 [28]. The field-contour orients more uniformly

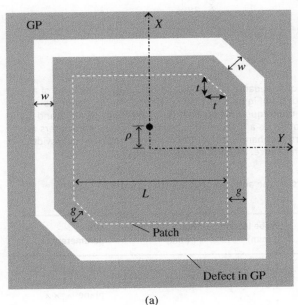

(a)

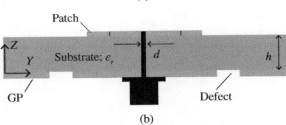

(b)

Figure 8.25 Schematic diagram of a coax-fed truncated corner square patch with ring-DGS for improved CP radiation [25]: (a) viewed from ground plane side with dotted lines indicating the patch boundary and white area as the etched-out DGS; (b) cross sectional view [25]. Source: Pasha et al. [25] © IEEE.

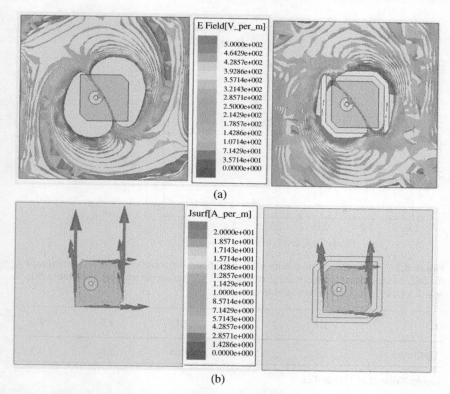

Figure 8.26 Simulated portrays for the antenna in Figure 8.25 with and without DGS [25]: (a) substrate electric-fields; (b) surface current on patch surface. Both are captured at 6.27 GHz for identical excitation phase. Parameters: $L = 14.22$ mm, $\rho = 2.5$ mm, $w = 1.5$ mm, $t = 2.1$ mm, $g = 1.65$ mm, $h = 1.575$, $\varepsilon_r = 2.33$, ground plane: 50 mm × 50 mm [25]. Source: Pasha et al. [25] © IEEE.

in presence of the DGS. The instantaneous surface current on a conventional patch surface exhibits both horizontal and vertical components although, ideally it should be unidirectional at a given instant of time. The unwanted orthogonal current is the source of XP radiation. It gets considerably reduced in presence of the DGS leading to an increase in polarization purity.

The outcome of the DGS-based CP design is examined in Figure 8.27 as a function of the truncations parameter t. A considerable improvement is revealed that occurs around a specific t value. For $t = 2.1$ mm, the traditional design shows 2.7 dB AR value and that with DGS reaches 0.2 dB. The improvement in AR is associated with the enhancement of effective operating bandwidth from 60 to 105 MHz and the reason behind this will be clear from Figure 8.28a. It is self-explanatory. The use of DGS hardly causes any change in S_{11} over the frequencies of interest, but a

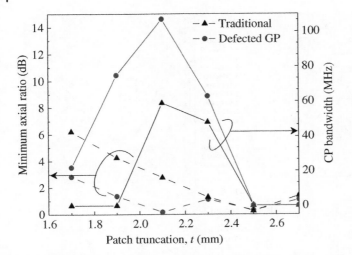

Figure 8.27 Variations in minimum AR and CP bandwidth with truncation parameter *t* for DGS-based antenna (Figure 8.25) at 6 GHz [25]. Antenna parameters as in Table 8.1; $w = 1.5$ ($\approx 0.032\lambda_0$), all dimensions in mm.

t	1.7	1.9	2.1	2.3	2.5
Optimum *g*	1.3	1.5	1.65	1.8	3.7

Source: Pasha et al. [25] © IEEE.

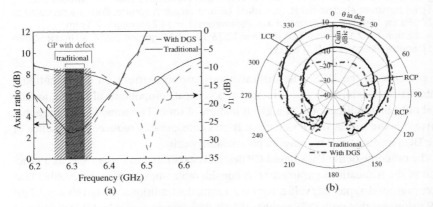

Figure 8.28 Input and radiation characteristics of the CP antenna (Figure 8.25) with and without DGS [25]; (a) Measured axial ratio and S_{11} versus frequency; (b) radiation patterns obtained at 6.3 GHz. Parameters as in Figure 8.26 with $t = 2.1$ mm [25]. Source: Pasha et al. [25] © IEEE.

remarkable widening in AR bandwidth is achieved. The measured radiation patterns at the frequency with the minimum AR value are shown in Figure 8.28b. The co-polar radiation remains unchanged, whereas the DGS results in as much as 10 dB suppression in cross-polar values along the boresight.

8.5.4 DGS for Beam Squint Correction and Improved CP Quality

A different type of problem is observed in thick substrate wideband CP design [9, 26, 29]. A typical antenna configuration is shown in Figure 8.29 [9, 29] where the patch is etched on a suspended substrate and feed inductance due to the long vertical probe is compensated by a capacitive ring, which was originally conceived by Hall [30]. The liner slot on the patch tilted by α angle with respect to x-axis actually controls the surface current in a CP fashion.

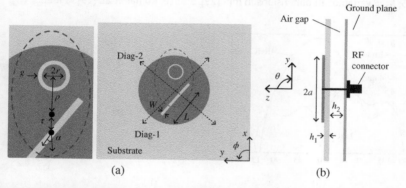

Figure 8.29 Wideband CP patch using suspended thick composite substrate [29]: (a) top view of a circular patch loaded with an inclined linear slot and an annular ring to compensate the feed inductance; (b) cross-sectional view using suspended substrate [29]. Source: Kumar et al. [29] © IEEE.

The change in the overall thickness $h_1 + h_2$ of the substrate of Figure 8.29 influences both AR and matching bandwidths as shown in Figure 8.30. They increase with increasing $h_1 + h_2$ and about 5% CP bandwidth is promised by $h_1 + h_2 \approx 0.11\lambda_0$ [29]. But larger thickness is associated with high XP radiation along with higher degree of beam squinting. Typical radiation performance of the antenna with 5% CP bandwidth is shown in Figure 8.31. This indicates its XP level as high as 0 dBic and a squint angle of the order of 16°. These appear as serious concerns as was also noticed in [26]. Thus, it is apparently difficult to utilize the advantage of a thick substrate in CP design.

A unique solution has been demonstrated in [29] that employed a DGS to rectify the source of the beam squinting. An asymmetry in fringing field distribution

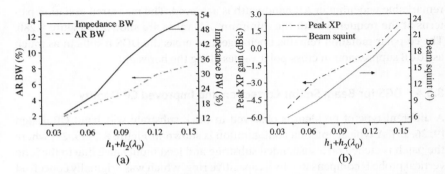

Figure 8.30 Characteristics of the CP antenna in Figure 8.29 as the function of substrate thickness $(h_1 + h_2)$: (a) AR and impedance bandwidth; (b) peak XP value and co-pol beam squint. Parameters: $a = 6.5$, $r = 1.9$, $g = 0.45$, $W = 0.65$, $L = 11.7$, $\rho = 2.7$, $\tau = 0.5$, $\alpha = 46°$, $h_1 = 0.5$, h_2 = variable; all dimensions in mm [29]. Source: Kumar et al. [29] © IEEE.

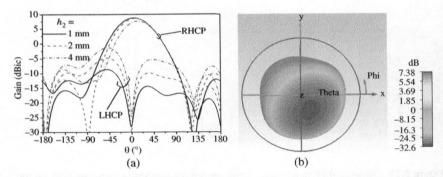

Figure 8.31 Radiation patterns of the RHCP patch as in Figure 8.29 [29]: (a) *xz*-plane radiation patterns for varying thickness $(h_1 + h_2)$; (b) simulated 3D radiation pattern at 8.2 GHz ($h_2 = 4$ mm) indicating beam squint. Other parameters as in Figure 8.30 [29]. Source: Kumar et al. [29] © IEEE.

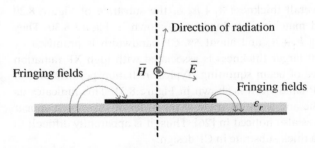

Figure 8.32 Effective near fields as a function of asymmetric fringe field distribution [29]. Source: Kumar et al. [29] © IEEE.

as depicted in Figure 8.32 is the real source which is primarily caused by the structural asymmetry as well as some degrees of spurious radiations from the long feeding probe [29]. The DGS is meant to create a new boundary condition to the fringing fields to compensate the said asymmetry but without disturbing the primary radiating fields. Three different configurations, as schematically shown in Figure 8.33 [29], were tested as annular ring (AL-DGS), symmetric arc (SA-DGS), and asymmetric arc (AA-DGS).

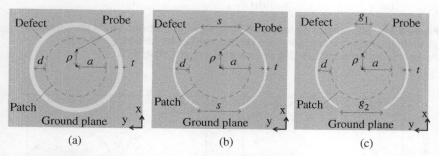

Figure 8.33 DGS geometries tested for correcting the beam squint: (a) AL-DGS, (b) SA-DGS, and (c) AA-DGS. Dotted line indicates patch boundary. $a = 6.5$ mm, $d = 3.4$ mm, $t = 1.1$ mm $h_1 = 0.5$ mm, $h_2 = 4$ mm (for AL-DGS), 0.7 mm (for SA-DGS, AA-DGS), $s = 7$ mm, $g_1 = 3.5$ mm, $g_2 = 9$ mm; defect: inner radius $= (a + d)$, width $= t$ [29]. Source: Kumar et al. [29] © IEEE.

Their impact in changing the beam squint scenario can be visualized from the 3D radiation patterns portrayed in Figure 8.34. The conventional ground plane without DGS shows about 16° squint angle over a diagonal plane and that reduces to only 7° with AL-DGS and 8° with AA-DGS. The SA-DGS reveals about 9° [29]. Thus, more than 50% correction in squint angle is possible by applying DGS. The measured radiations with AA-DDG are depicted in Figure 8.35. Another noticeable feature is the reduction in cross-polarized LHCP radiation which is of the order of 2–6 dB over two representative planes covering up to about 60° elevation angle.

The physical change into the fringing fields owing to the DGS is clearly evident from Figure 8.36. Unbalanced and nonuniform distribution across *xz*-cut of a conventional geometry remarkably transforms more balanced and uniform nature in presence of the DGS. This is the operating principle of this DGS-based improvement.

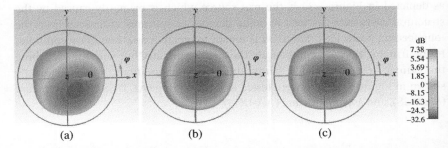

Figure 8.34 3D radiation patterns at 8.2 GHz showing the beam squint scenario under different ground plane configurations [29]: (a) no-DGS, (b) AL-DGS, (c) AA-DGS. Parameters as in Figure 8.33 [29]. Source: Kumar et al. [29] © IEEE.

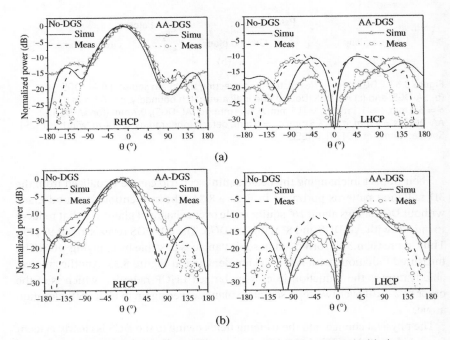

Figure 8.35 Measured radiations with and without AA-DGS compared with the simulated predictions ($f = 8.2$ GHz) [29]: (a) yz-plane; (b) diagonal plane (dig-1 in Figure 8.29). Parameters as in Figure 8.33 [29]. Source: Kumar et al. [29] © IEEE.

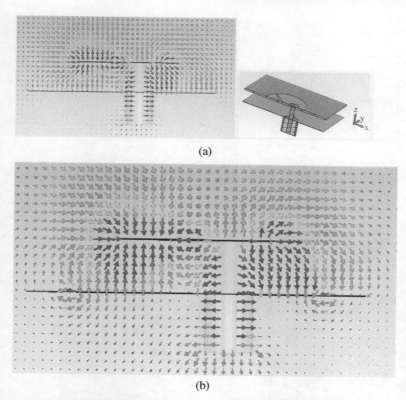

(a)

(b)

Figure 8.36 Simulated fringing fields obtained at 8.2 GHz across *xz*-plane [29]: (a) without any DGS; (b) with AL-DGS. Parameters as in Figure 8.33. [29]. Source: Kumar et al. [29] © IEEE.

8.6 Evolving Applications: DGS in SIW-Based CP Antennas

Like microstrip, substrate integrated waveguide (SIW) technology has also successfully explored DGS for antenna applications. Figure 8.37 is a good example [31] representing a dual circularly polarized antenna. It comprises an X-band quarter mode SIW cavity loaded with a couple of open stubs and a combination of multiple DGSs. The stubs help in adjusting the dual resonances and also in enhancing the CP bandwidth.

The DGS assembly is strategically deployed on the back of the cavity to realize high isolation between the feeding ports. An interesting experimentation with the shape, size, number, and orientation of the DGSs is depicted through Figure 8.38.

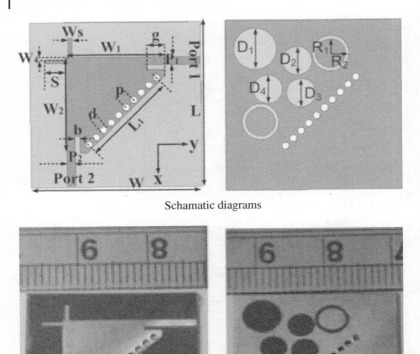

Schamatic diagrams

Photographs of a prototype

Figure 8.37 SIW-based dual CP antenna with DGS [31]. Left side; top view: SIW Cavity with open stubs, Right side; bottom view: combination of DGSs. Source: Kumar et al. [31], IEEE.

The sequence of the changes in geometries is sketched in Figure 8.38a and corresponding changes of port isolation in Figure 8.38b. It records an improvement by about 9 dB for Antenna #4. An equally important result is furnished in Figure 8.38c which reveals a distinct improvement in AR bandwidth (Antennas #1 to #4) as a contribution of the DGS. Possibility of improving it by more than 20% has been ensured.

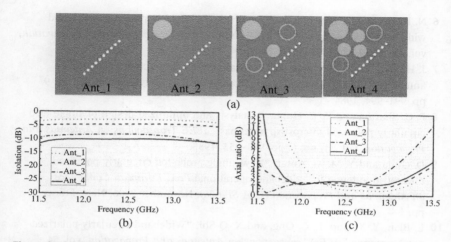

Figure 8.38 DGS configuration and their performances with reference to the design in Figure 8.37 [31]: (a) Evolution of DGS configuration; (b) isolation between port 1 and port 2; (c) axial ratio characteristics [31]. Source: Kumar et al. [31] © IEEE.

References

1 T. Manabe, K. Sato, H. Masuzawa, K. Taira, T. Ihara, Y. Kasashima, and K. Yamaki, "Polarization dependence of multipath propagation and high-speed transmission characteristics of indoor millimeter-wave channel at 60 GHz," *IEEE Transactions on Vehicular Technology,* vol. 44, no. 2, pp. 268–274, 1995.

2 C. A. Balanis, *"Antenna Theory Analysis and Design",* Wiley, Hoboken, New Jersey, 2005.

3 S. Ogurtsov and S. Koziel, "Approach to axial ratio improvement for circularly polarized microstrip patch antennas excited via two inputs," *IET Microwaves, Antennas & Propagation,* vol. 10, no. 7, pp. 770–776, 2016.

4 P. S. Hall, J. S. Dahele, and J. R. James, "Design principles of sequentially fed, wide bandwidth, circularly polarized microstrip antennas," *IEE Proceedings – Microwaves, Antennas and Propagation,* vol. 136, no. 5, pp. 381–389, 1989.

5 J. Huang, "A technique for an array to generate circular polarization with linearly polarized elements," *IEEE Transactions on Antennas and Propagation,* vol. 34, no. 9, pp. 1113–1124, 1986.

6 N. Herscovici, Z. Sipus, and D. Bonefacic, "Circularly polarized single-fed wide-band microstrip patch," *IEEE Transactions on Antennas and Propagation*, vol. 51, no. 6, pp. 1277–1280, 2003.

7 J. Huang, "Circularly polarized conical patterns from circular microstrip antennas," *IEEE Transactions on Antennas and Propagation*, vol. 32, no. 9 pp. 991–994, 1984.

8 P. C. Sharma and K. C. Gupta, "Analysis and optimized design of single feed circularly polarized microstrip antennas," *IEEE Transactions on Antennas and Propagation*, vol. 31, no. 6, pp. 949–955, 1983.

9 D. Guha and Y. M. M. Antar, "New single probe-fed circularly polarized microstrip antenna for wideband operation," *Proc. European Conference on Antennas and Propagation (ESA SP-626)*, November 2006, Nice, France, pp. 1–4.

10 L. Bian, Y. X. Guo, L. C. Ong, and X. Q Shi, "Wideband circularly-polarized patch antenna," *IEEE Transactions on Antennas and Propagation*, vol. 54, no. 9, pp. 2682–2686, 2006.

11 K. R. Carver and J. W. Mink, "Microstrip antenna technology," *IEEE Transactions on Antennas and Propagation*, vol. 29, no. 1, pp. 2–24, 1981.

12 D. Guha, M. Biswas, and Y. M. M. Antar, "Microstrip patch antenna with defected ground structure for cross polarization suppression," *IEEE Antennas and Wireless Propagation Letters*, vol. 4, pp. 455–458, 2005.

13 X. L. Bao and M. J. Ammann, "Dual frequency circularly polarized patch antenna with compact size and small frequency ratio," *IEEE Transactions on Antennas and Propagation*, vol. 55, no. 07, pp. 2104–2107, 2007.

14 K. Wei, J. Y. Li, L. Wang, R. Xu, and Z. J. Xing, "A new technique to design circularly polarized microstrip antenna by fractal defected ground structure," *IEEE Transactions on Antennas and Propagation*, vol. 65, no. 07, pp. 3721–3725, 2017.

15 K. Wei, B. Zhu, and M. Tao, "The circular polarization diversity antennas achieved by a fractal defected ground structure," *IEEE Access*, vol. 7, pp. 92030–92036, 2019.

16 Y. Zhang, Z. Han, S. Shen, C. Y. Chiu, and R. Murch, "Polarization enhancement of microstrip antennas by asymmetric and symmetric grid defected ground structures," *IEEE Open Journal of Antennas and Propagation*, vol. 1, pp. 215–223, 2020.

17 B. Kim, B. Pan, S. Nikolaou, Y. S. Kim, J. Papapolymerou, and M. M. Tentzeris, "A novel single fed circular microstrip antenna with reconfigurable polarization capability," *IEEE Transactions on Antennas and Propagation*, vol. 56, no. 3, pp. 630–638, 2008.

18 W. S. Yoon, J. W. Baik, H. S. Lee, S. Pyo, S. M. Han, and Y. S. Kim, "A reconfigurable circularly polarized microstrip antenna with a slotted ground plane," *IEEE Antennas and Wireless Propagation Letters*, vol. 9, pp. 1161–1164, 2010.

19 X. X. Yang, B. C. Shao, F. Yang, A. Z. Elsherbeni, and B. Gong, "A polarization reconfigurable patch antenna with loop slots on the ground plane," *IEEE Antennas and Wireless Propagation Letters*, vol. 11, pp. 69–72, 2012.

20 S. J. Lee, H. J. Lee, W. S. Yoon, S. J. Park, J. Lim, D. Ahn, and S. M. Han, "Circular polarized antenna with controlled current distribution by defected ground structures," In *Proceedings of International Symposium on Antennas and Propagation (ISAP)*, pp. 1–4, 2010.

21 P. R. Prajapati, G. G. K. Murthy, A. Patnaik, and M. V. Kartikeyan, "Design and testing of a compact circularly polarised microstrip antenna with fractal defected ground structure for L-band applications," *IET Microwaves, Antennas and Propagation*, vol. 9, no. 11, pp. 1179–1185, 2015.

22 J. P. Thakur and J. S. Park, "An advance design approach for circular polarization of the microstrip antenna with unbalance DGS feedlines," *IEEE Antennas and Wireless Propagation Letters*, vol. 5, pp. 101–103, 2006.

23 Y. X. Guo, K. W. Khoo, and L. C. Ong, "Wideband circularly-polarized patch antenna using broadband baluns," *IEEE Transactions on Antennas and Propagation*, vol. 56, no. 2, pp. 319–326, 2008.

24 Y. X. Guo, Z. Y. Zhang, and L. C. Ong, "Improved wide-band Schiffman phase shifter," *IEEE Transactions on Microwave Theory and Techniques*, vol. 55, no. 3, pp. 1196–1200, 2006.

25 M. I. Pasha, C. Kumar, and D. Guha, "Application-friendly improved designs of single-fed circularly polarized microstrip antenna," *IEEE Antennas and Propagation Magazine*, vol. 61, no 03 pp. 80–89, 2019.

26 J. M. Kovitz and Y. R. Samii, "Using thick substrates and capacitive probe compensation to enhance the bandwidth of traditional CP patch antennas," *IEEE Transactions on Antennas and Propagation*, vol. 62, no. 10, pp. 4970–4979, 2014.

27 M. I. Pasha, C. Kumar, and D. Guha, "Investigations into improved isolation in co- to cross-polar radiation fields of a single-fed circularly polarized patch antenna," *Proceedings of INDICON*, pp. 1–4, December 15–17th, 2016, Bangalore.

28 High frequency structure simulator (HFSS), Ansoft, v 11.1

29 B. P. Kumar, D. Guha, and C. Kumar, "Reduction of beam squinting and cross-polarized fields in a wideband CP element," *IEEE Antennas and Wireless Propagation Letters*, vol. 19, no. 3, pp. 418–422, 2020.

30 P. S. Hall, "Probe compensation in thick microstrip patches," *Electronics Letters*, vol. 23, no. 11, 1987, pp. 606–607 (DOI: https://doi.org/10.1049/el: 19870434).

31 K. Kumar, S. Dwari, and M. K. Mandal, "Broadband dual circularly polarized substrate integrated waveguide antenna," *IEEE Antennas and Wireless Propagation Letters*, vol. 16, pp. 2971–2974, 2017.

9

DGS Integrated Printed UWB Monopole Antennas

9.1 Introduction

The ultrawideband (UWB) technology was conceived as early as the late 1960s. The primary development was for military applications using transceiver systems with extremely wide operating bandwidth. It gained momentum after the release of the first report and order [1] by the Federal Communications Commission (FCC). This allowed the unlicensed use of a band of frequencies covering 3.1–10.6 GHz for commercial purpose but with EIRP (effective isotropic radiated power) less than −41.3 dBm/MHz to avoid interference with the other communication systems. It encouraged short-range UWB communications between portable electronic devices and also created extra interest among the wireless engineer in developing UWB enabled devices.

Planar monopole appears to be a highly suitable candidate for such UWB operation and as a result an extensive research on this has been recorded over the years [2]. Design of UWB planar monopole with consistent gain over the entire bandwidth maintaining a linear phase response is truly challenging. Another challenge is the restriction on the ground plane size which naturally degrades the antenna performance in terms of both impedance matching and uniform radiations. In addition, the impedance characteristics of such UWB antenna is supposed to take care of a few narrow intermediate bands which are allocated to other commercial services such as WLAN (wireless local area network) operating over 5.15–5.35 GHz/5.725–5.825 GHz and WiMAX (Worldwide Interoperability for Microwave Access) operating over 3.3–3.6 GHz. The UWB antenna should not be disturbed by those frequencies and vice versa, and therefore, adequate measures has to be taken in the antenna design.

The printed monopole with engineered ground of defects features stopband characteristics and that mechanism became gradually popular since around 2007 [3]. Apart from that feature, DGS-based design facilitates some additional advantages such as achieving (i) large impedance bandwidth with compact

Defected Ground Structure (DGS) Based Antennas: Design Physics, Engineering, and Applications,
First Edition. Debatosh Guha, Chandrakanta Kumar, and Sujoy Biswas.

ground plane, (ii) multiband operations, (iii) band notched characteristics to avoid usual interference, and (iv) improving isolation between closely packed elements in MIMO applications.

9.2 Improved Impedance Bandwidth and Multiband Operation

The concept of using slots of different geometries on planar monopoles was reported in the early 2000s to realize band rejection characteristic [2] within its wide operating bandwidth. As mentioned earlier, such band rejection is required to avoid interference between UWB and other wireless bands. Placing such slots or slits as defects on the ground plane serves the purposes successfully.

9.2.1 Improved Impedance Matching of UWB Antennas

A planar monopole itself is a UWB antenna, but challenges come when its ground plane size is restricted in view of portability of the device. Thus, obtaining impedance matching over a wide operating band needs some additional efforts. Different solutions are discussed in [2]. An improved design in [4] employed ground plane defects as shown in Figure 9.1a. The monopole is of rectangular shape etched on FR4 substrate. A pair of narrow notches measuring $L_1 \times W_1$ (on the monopole) and $L_2 \times W_2$ (on the ground plane) has been created to control the electromagnetic coupling between the monopole and its ground in order to achieve the required impedance matching. Figure 9.1b shows the measured results revealing about 112% impedance bandwidth covering 3.1–11 GHz [4]. Similar kind of ground plane notches of various shapes has been explored in UWB monopole designs [5–9]. The geometry in Figure 9.1a [4] has been modified in [10] and this looks as in Figure 9.2 [10]. The purpose is the same and this design claims a matching bandwidth from 3.1 to 11.8 GHz.

A new approach was demonstrated in [11] to obtain even wider impedance bandwidth from 2.95 to 12.81 GHz. The primary radiator is a CPW fed fractal-shaped monopole as shown in Figure 9.3 [11]. The ground plane geometry is quite interesting. It comprises rectangular defects along with notches in a balanced configuration. An innovation in the defect geometry is shown in Figure 9.4 [12] in view of accommodating them in the limited sized ground plane. It originates from a fractal and effectively increases the current path within the available space on the ground plane [12]. The defect geometry in Figure 9.4 also balances the distribution of horizontal versus vertical conduction currents to avoid any nonuniformity in the radiation patterns.

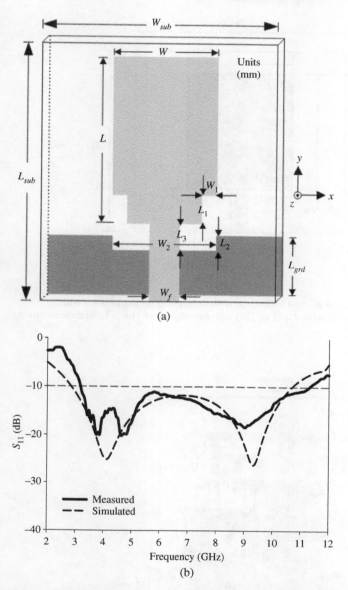

(a)

(b)

Figure 9.1 (a) Printed monopole with engineered ground plane for the required impedance matching [4]: (a) antenna layout; (b) Simulated and measured S_{11} versus frequency. Optimized dimensions: $W_{sub} = 16$, $L_{sub} = 18$, $W = 7$, $L = 11$, $W_1 = 1$, $L_1 = 2$, $W_2 = 7$, $L_2 = 1$, $L_3 = 3$, $W_f = 2$, $L_{grd} = 4$, substrate thickness = 1.6 with $\varepsilon_r = 4.4$, all dimensions in mm. Source: Jung et al. [4] IEEE.

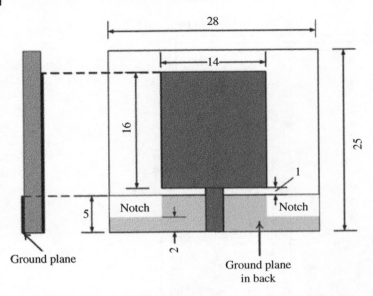

Figure 9.2 Rectangular printed monopole with defected ground plane for wide matching bandwidth. Source: Wu et al. [10] with permission of The IET. All dimensions on the figure are in mm.

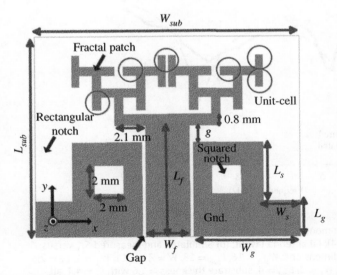

Figure 9.3 CPW fed fractal monopole antenna with defected ground plane. Source: Jalali and Sedghi [11] John Wiley & Sons.

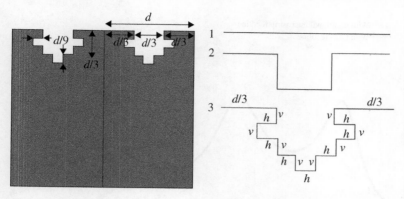

Figure 9.4 Fractal shaped defect on the ground plane and the origin of the fractal geometry. Source: Fereidoony et al. [12] IEEE.

A further improvement was reported in [13] and that antenna geometry is shown in Figure 9.5. It employs a pair of U-shaped DGSs beneath the feed along with a tuning stub on the feed line. The role of DGS is to control input impedances near two adjacent frequencies as examined in Figure 9.6 [13]. This is manifested through an increase in bandwidth from 41.9% (without DGS) to 90.4% (with DGS)

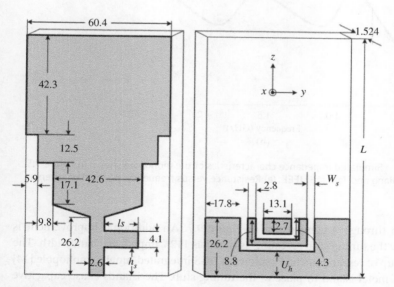

Figure 9.5 Trapezoidal printed monopole integrated with double U-shaped DGS and an open stub. All dimensions on the figure are in mm. Source: Chiang and Tam [13] IEEE.

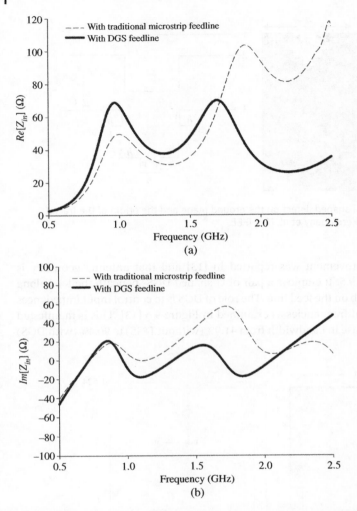

Figure 9.6 Simulated impedance characteristics of the monopole shown in Figure 9.5. Source: Chiang and Tam [13] IEEE: (a) Resistance versus frequency; (b) reactance versus frequency.

as shown through a comparison in Figure 9.7. An additional improvement is caused by the tuning stub revealing as much as 112% impedance bandwidth. The same group of researchers had explored a DGS integrated tunable monopole [14] and used metal island in place of the tuning stub. The islands cause capacitive loading and accommodate varactor diodes for tuning purpose. It promises a tuning range from 2.7 to 2.1 GHz.

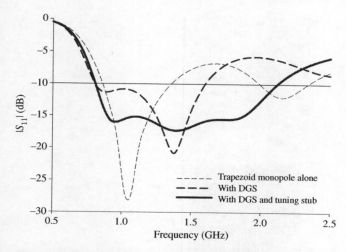

Figure 9.7 Simulated S_{11} versus frequency for the monopole in Figure 9.5 with/without DGS and tuning stub. Source: Chiang and Tam [13] IEEE.

9.2.2 DGS Induced Resonances for Improved UWB Operation

The resonance occurring in a DGS slot can be utilized for improved performance in terms of bandwidth, multiband operation, and radiations of a monopole [15, 16]. The idea is simple as shown in Figure 9.8 which is meant for multiband design. It is a CPW feed circular head monopole. The circular geometry is straight forward and provides sufficiently large impedance bandwidth. A reference antenna without DGS and using a considerably large ground plane has been studied in Figure 9.9 indicating its wide operating band that starts from 3.06 GHz. Such monopoles are generally used in high-speed mobile devices and MIMO arrays. But the challenge is to realize it on a small ground plane as elaborately demonstrated in [16]. Introduction of an L-shaped DGS perturbs the surface current as documented in Figure 9.10. Its impact is revealed by examining the prototypes with and without DGS (Figure 9.11).

Their impedance characteristics are compared in Figure 9.12. The prototype without DGS (Figure 9.12a) exhibits a wide bandwidth from about 4–10 GHz (Figure 9.12a). The DGS integrated version (Figure 9.11b) reveals two operating bands across 2.68–3.28 GHz and 4.74–9.58 GHz (Figure 9.12b) of which the lower band is produced by the DGS. The radiations patterns discussed in [16] ensure the presence of two orthogonal polarizations at 2.7 GHz and dominantly unipolar feature around 5.5 GHz.

The L-shaped defect on the ground has been tried later in combination with transmission line-metamaterial (TL-MTM) to generate a compact triband

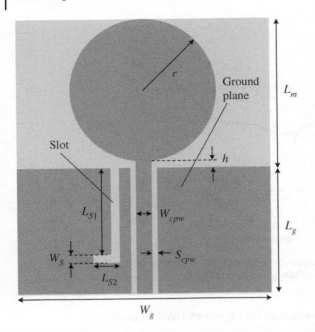

Figure 9.8 CPW fed circular head printed monopole with a single L-shaped DGS [16]. Parameters: $r = 8$, $h = 0.3$, $L_m = 16.3$, $W_g = 24$, $L_g = 12$, $W_{cpw} = 1.55$, $S_{cpw} = 0.2$, $W_s = 1$, $L_{s1} = 9$, $L_{s2} = 3$, substrate thickness = 1.59 with $\varepsilon_r = 4.34$, all dimensions in mm. Source: Antoniades and Eleftheriades [16] IEEE.

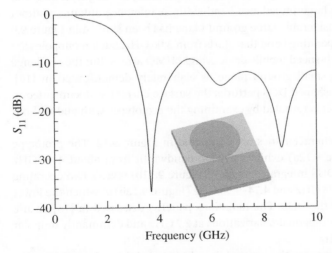

Figure 9.9 Simulated S_{11} of a CPW-fed circular monopole (Figure 9.8) without DGS. Ground plane size $W_g = 40$ mm, $L_g = 20$ mm, other parameters as in Figure 9.8.

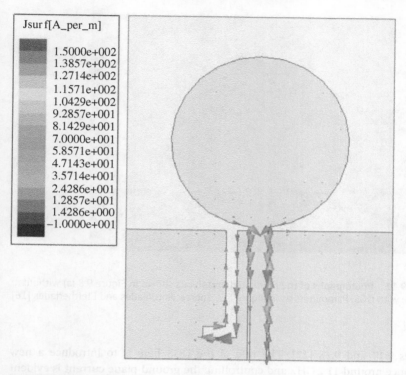

Figure 9.10 Simulated surface current on the DGS integrated monopole at 2.7 GHz [16]. Parameters as in Figure 9.8. Source: Adapted from Antoniades and Eleftheriades [16].

monopole as shown in Figure 9.13 [17]. The unit cell TL-MTM comprises a pair of rectangular patches C_1 and C_2 connected by a narrow strip L_2 which offers negative refractive index (NRI). The other side of the ground plane embodies a single L-shaped defect. The air bridge ensures balanced current flow through the CPW ground which is compact in size (20.0 mm × 23.5 mm × 1.59 mm). The step-wise development of the antenna is nicely depicted through Figure 9.14. A conventional geometry (Case I) offers a single resonance around 6 GHz. Case II, i.e. a single TL-MTM cell-loaded monopole follows the concept of [18, 19] and produces dual resonances. The final form is Case III which adds an L-DGS and results in triple band performance to serve two WiFi bands (2.40–2.48 GHz; 5.15–5.80 GHz) and the WiMAX service from 3.30–3.80 GHz. The prototype of the antenna is shown in Figure 9.15 along with its measured S_{11}.

Balanced L-DGS has also been explored in similar monopole designs [20, 21]. Figure 9.16 [20] shows one such example which also uses a parasitic loading. Single T-shaped defect also works in a very interesting way as shown through

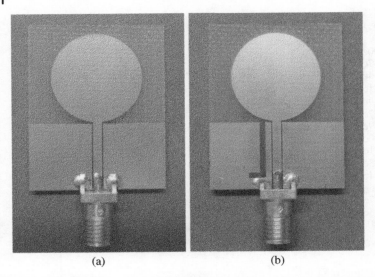

(a)   (b)

Figure 9.11 Photographs of the fabricated prototypes shown in Figure 9.8 (a) without DGS; (b) with DGS. Parameters as in Figure 9.9. Source: Antoniades and Eleftheriades [16] IEEE.

Figures 9.17 and 9.18 [22]. The role of the DGS here is to introduce a new resonance around 11.2 GHz and controlling the ground plane current is evident from the studies in Figure 9.18. This eventually results in a purely UWB operation over 3.12–12.73 GHz. A remarkable improvement has also been recorded using V-shaped DGS in [23].

9.3 Band Notch Characteristics in UWB Antennas

One of the most widely used applications of DGS is to obtain single or multiple stopband notches within the UWB characteristics of planar monopoles. The purpose is to eliminate the possibility of interference with some other application frequencies which fall within the allocated UWB band. Most of them spread below 6 GHz and they are addressed in 9.3.1. A few designs also extend up to 10 GHz as discussed in 9.3.2.

9.3.1 DGS Based UWB Antenna to Avoid Interference up to C-Band

Several wireless applications are widely populated around 2–6 GHz which commonly include Bluetooth/WLAN (2.4/5.2/5.8 GHz), WCDMA (2.1 GHz), and

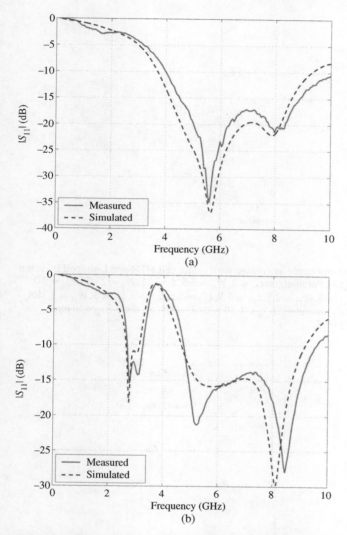

Figure 9.12 S_{11} characteristics of the prototypes shown in Figure 9.11 [16]: (a) without DGS; (b) with DGS. Source: Antoniades and Eleftheriades [16] IEEE.

WiMAX (3.5/5.5 GHz). The antenna researchers, therefore, paid more attention to resolving this aspect by different techniques [2] where most of them included slots on the radiator or the feed. A DGS was employed in [3] showing a band notch characteristic of a CPW fed UWB rhombic monopole. The geometry is shown in Figure 9.19 [3] and one should be careful in understanding its

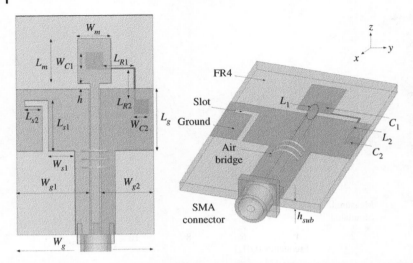

Figure 9.13 CPW fed monopole integrated with single cell MTM and L-shaped DGS: top and isometric views [17]. Parameters: $L_m = 8$, $W_m = 5.8$, $L_g = 11$, $W_g = 23.5$, $W_{g1} = 9.0$, $W_{g2} = 12.5$, $h = 1$, $W_{c1} = 3$, $W_{c2} = 2.5$, $L_{R1} = 5.5$, $L_{R2} = 5.5$, $L_{s1} = 9$, $L_{s2} = 3$, $W_{s1} = 7$, slot width $g = 1$. Substrate parameters: $h_{sub} = 1.59$ with $\varepsilon_r = 4.4$, all dimensions in mm. Source: Zhu et al. [17] IEEE.

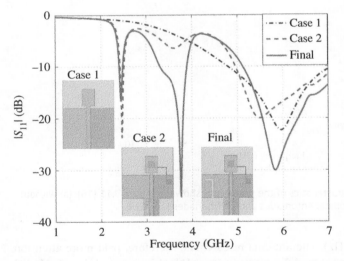

Figure 9.14 Simulated S_{11} characteristics for three designs: Case 1- reference antenna (unloaded monopole); Case 2- monopole loaded with single cell MTM; Final Case- monopole with single cell MTM and L-shaped DGS as in Figure 9.13. Parameters as in Figure 9.13. Source: Zhu et al. [17] IEEE.

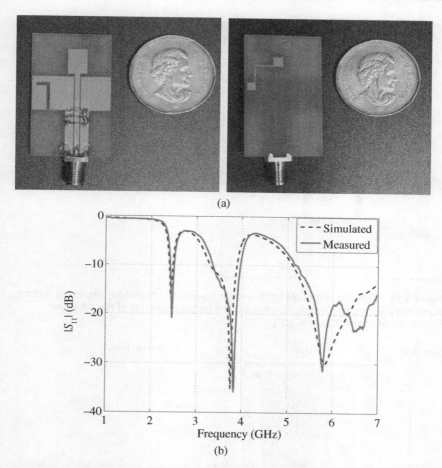

(a)

(b)

Figure 9.15 Single MTM cell and the DGS loaded planar monopole [17]: (a) Photographs of a prototype; (b) S_{11} characteristics indicating tri-band operation Source: Zhu et al. [17] IEEE.

design strategy. Two symmetrical notches in the ground plane enhances the impedance bandwidth following the principle discussed in Section 9.2.1. A straight slit measuring $L_2 \times W_2$ produces band-rejection from 5.1 to 5.81 GHz. The slit dimensions such as L_2 ($\approx \lambda_g/4$) and W_2 appear respectively important in controlling the rejection frequency and rejection band. Similar slit and notch combination has been successfully explored for band rejection purpose in various forms [6, 24–28]. Figure 9.20 depicts a set of representative designs and their aims in achieving a feature of dual rejection frequency. One such example is shown

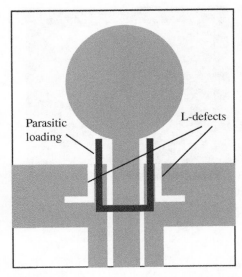

Figure 9.16 CPW fed monopole antenna with balanced L-shaped defects on the ground plane and a U-shaped parasitic element printed on the other side of the substrate. Source: Adapted from Liu et al. [20].

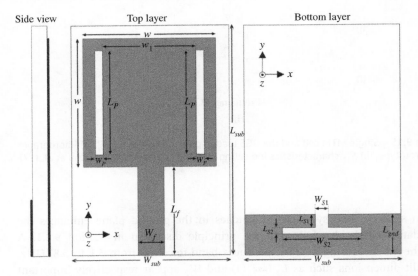

Figure 9.17 Slot loaded square patch monopole with T-shaped DGS. Source: Ojaroudi et al. [22] IEEE.

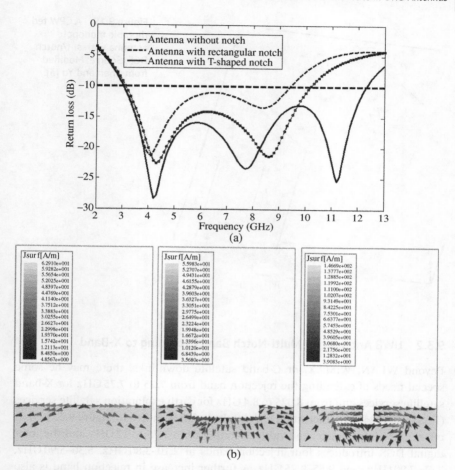

Figure 9.18 Influence of the defects on the monopole design in Figure 9.17 [22]: (a) S_{11} characteristics indicating contributions of the DGS; (b) Ground plane current distribution at 11.2 GHz for conventional and notched ground planes. Source: Ojaroudi et al. [22] IEEE.

in Figure 9.21 which indeed characterizes the antenna depicted in Figure 9.20c. The VSWR ensures two rejection bands working from 3.3 to 3.8 GHz and 5.1 to 6 GHz respectively. The simulated surface currents help in identifying the reasons behind them. The slots on the monopole surface control the first rejection frequency whereas the DGS is responsible for the second one.

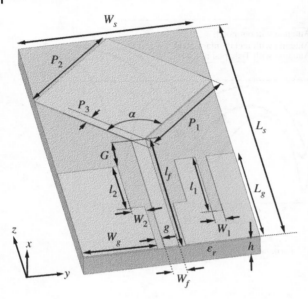

Figure 9.19 A CPW fed rhombic monopole antenna with slit/notch DGS. Source: Modified from Chen and Yu [3].

9.3.2 UWB Antenna for Multi-Notch Band Extending to X-Band

Beyond WLAN, WiMAX, or C-band satellite downlinks, there may be some special needs of extending the rejection band from 7.25 to 7.75 GHz for X-band satellite services and from 8.025 to 8.4 GHz for earth exploration satellite services (EESSs) [29–34]. An interesting but simple design is shown in Figure 9.22 [29]. It offers 140% impedance bandwidth operating from 1.75 to 12 GHz and the inter digital DGS introduces four rejection bands at 2.10–3.80 GHz, 5.30–5.90 GHz, 7.65–7.90 GHz, and 9.55–9.75 GHz. A further increase in rejection band is also possible. An investigation in [30] has reported such results using additional engineering on the monopole surface.

The challenge in such UWB antenna design is to critically adjust all rejection bands simultaneously maintaining individual bandwidths. It becomes more and more difficult as the number of rejection frequencies increases. An interesting technique has been explored in [34] which uses DGS along with step impedance resonator slot. Numerous band notched UWB antennas have been tried with different element cum DGS geometries [5, 7, 8, 35–41]. The basic design approach is identical but the features vary with their band of operation and mode of applications. A comparative study is presented in Table 9.1.

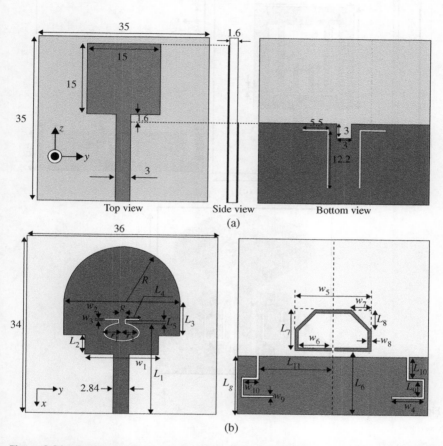

Figure 9.20 Schematic diagrams of a few representative UWB monopoles with dual DGS units: (a) microstrip line-fed square patch monopole with dual inverted L-shaped DGS. Source: Jiang et al. [6] John Wiley & Sons; (b) printed fat monopole loaded by an open loop parasitic element and Ω shaped slot on its body. Source: Li et al. [24] IEEE; (c) Stub loaded patch monopole with an additional modified G-shaped DGS underneath the feed line. Source: Abdollahvand et al. [25] IEEE; (d) CPW fed fan shaped monopole with a trapezoidal ground plane bearing double U-shaped DGS. Source: Li et al. [26] John Wiley & Sons.

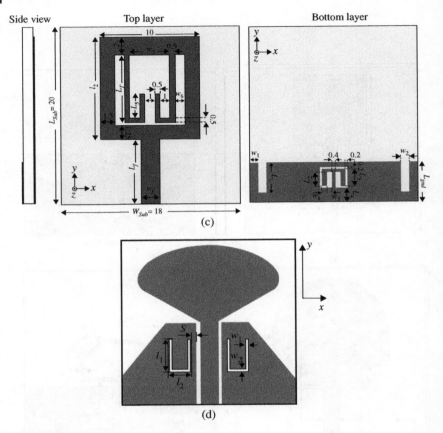

Figure 9.20 *(Continued)*

9.4 Applications to Band Notch UWB MIMO Antennas

UWB MIMO antenna is also important [42–44] and one such example is shown in Figure 9.23 [42]. A pair of orthogonally oriented stepped slot antennas have been designed to operate from 3.5 to 10.6 GHz. Three different DGS shapes have been employed purposefully to introduce two notch bands around 5.6 and 8 GHz (by L-shaped and U-shaped DGSs respectively) and greater than 15 dB port isolation (by Y-shaped DGS).

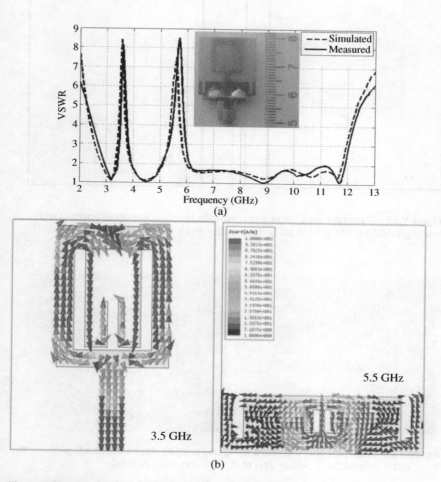

Figure 9.21 Characterization of the antenna structure shown in Figure 9.20c [25]: (a) measured and simulated VSWR versus frequency; (b) Simulated current portrays on the patch surface at 3.5 GHz and ground plane at 5.5 GHz. Source: Abdollahvand et al. [25] IEEE.

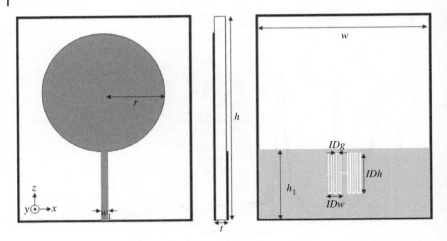

Figure 9.22 Schematic diagram of a microstrip fed circular monopole antenna with interdigital DGS in the ground plane. Source: Lee and Kim [29] John Wiley & Sons.

Figure 9.23 Dual port MIMO antenna using two stepped slot antennas integrated with Y-, L-, and U-shaped DGSs. Source: Zhu et al. [42] John Wiley & Sons.

An interesting 4-port MIMO design is shown in Figure 9.24 [44]. The antenna layout clearly indicates four quadrants. Two quadrants are used to deploy two UWB elements on one face of the substrate (say #1 and #3 on the top side) and the other two elements are placed on the reverse face occupying the rest of the quadrants (say #2 and #4). The dotted lines represent the layout on the reverse face. Both L- and U-shaped defects have been used for individual ground plane which result in stopbands around 3.5 and 7.5 GHz respectively. The shaped slot provides one more stopband in between these two.

Table 9.1 DGS based band notch UWB antennas: comparisons of a few representative designs.

References	Area (mm²)	Operating frequency (GHz)	Impedance bandwidth %BW	Rejection frequency (GHz)	Rejection band(s)	DGS types
[3]	17×28.5	3.1–11.9	118	5.1–5.81	Single	DGS slots
[5]	18×30	3–16	136	5–6	Single	DGS slots
[6]	35×35	2.8–10.7	117	4.8–57	Single	Dual inverted L DGS
[24]	34×36	2.9–13	127	3.3–3.9, 5.2–5.35, 5.8–6.0	Triple	Dual Hook shaped DGS, slot in radiator, metal open ring
[25]	18×20	2.8–11.8	123	3.3–3.8, 5.1–6	Dual	I-shape and G-shape DGS, and stubs in the radiator
[26]	30×30	3–15	133	5–6	Single	Dual U-shaped DGS
[27]	15×18	3.1–14	128	5.13–6.1	Single	Shovel shaped DGS
[28]	18.4×21.5	3–16	130	5.15–5.85	Single	Dual symmetric DGS slits
[29]	28×35	1.75–12	149	2.10–3.8, 5.3–5.9, 7.65–7.9, 9.55–9.75	Quadruple	Interdigital DGS
[30]	80×80	1.5–11	152	1.85–22.49, 2.91–3.40, 3.95–5.39, 5.65–6.25, 7.19–8.78	Five	Inverted U and open loop DGS arc slot and metal open loop resonator
[31]	21×28	3.1–10.6	109	5–6, 7.7–8.5	Dual	H-shaped DGS, double inverted U slot in radiator

(Continued)

Table 9.1 (Continued)

References	Area (mm²)	Operating frequency (GHz)	Impedance bandwidth %BW	Rejection frequency (GHz)	Rejection band(s)	DGS types
[32]	30×45	2.4–12	133	3.15–3.75, 4.85–6.08, 7.98–8.56	Triple	CMLSRR DGS and slot on radiator
[33]	24×42	3.1–11	112	3.3–3.7, 7.1–7.76, 5.15–5.825	Triple	Double U and extended U DGS, arc slot in radiator
[34]	21×28	2.8–11.3	120	3.1–3.9, 5–6.1, 7.2–7.78	Triple	H-shaped DGS, fork shaped stubs
[35]	22×22	3.1–14	127	5.1–5.9	Single	DGS slots
[36]	35×45	1.62–17.43	166	Around 5.8	Single	Quarter and half wavelength DGS slots
[37]	21×28	2.38–10.6	126	5–6	Single	L shaped DGS slots
[8]	24×25	3.05–14.2	129	5.14–5.36, 5.74–6.07	Dual	DGS slits and SRR
[7]	35×40	2.2–11	133	2.3–2.9, 5.5–6.3	Dual	Dual inverted L DGS and notch

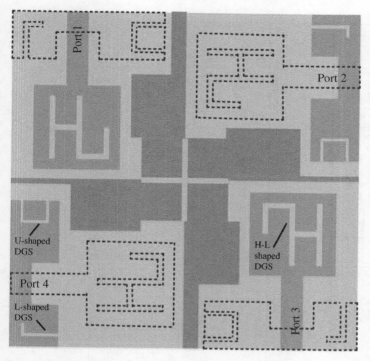

Figure 9.24 Layout of shaped slot loaded 4-square patches for a MIMO antenna design using L-, U-shaped defects on the ground plane. The dotted lines represent antennas and ground plane on reverse face of the substrate. Source: Tang et al. [44] IEEE.

9.5 Time Domain Behavior of DGS Based UWB Monopole

A wide range of DGS integrated UWB monopoles have been discussed above indicating their band notch characteristics. Their time domain behavior is equally important to be studied for high-speed data communications. Very limited investigations are available in the open literature and one representative study is documented in [45]. Figure 9.25a through d show three techniques such as radiator with elliptic slot, parasitic ring near feedline, and slot DGS, to achieve rejection notches around 5.8 GHz WLAN band with identical rejection feature. The received signals for those three configurations are shown in Figure 9.26 and all of them reveal prominent ringing with varying decaying property. The monopole with DGS (Figure 9.25d) indicates relatively fast decay with time and

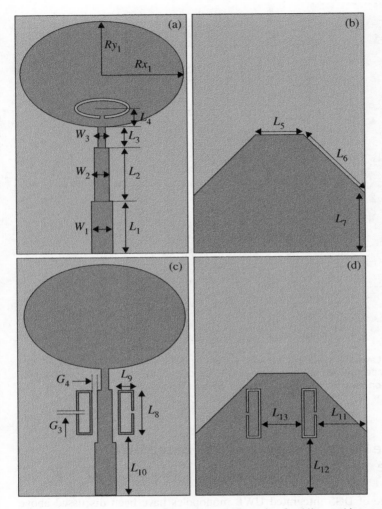

Figure 9.25 (a) UWB antenna feed with multi-section feed line and integrated with elliptic slot (b) Ground plane structure for the monopole; (c) UWB antenna with parasitic element around its feedline; (d) DGS integrated configuration. Source: Malik et al. [45] IEEE.

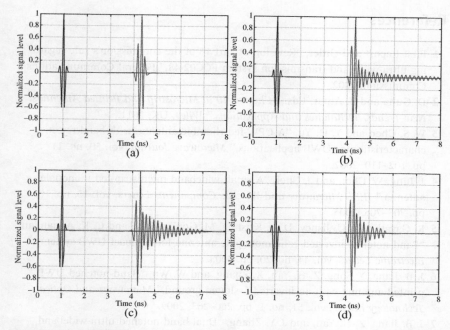

Figure 9.26 Time domain response (input pulse followed by the received signal) of UWB monopoles with and without DGS: (a) a typical reference UWB monopole without any DGS; (b) slot loaded monopole without DGS (Figure 9.25a); (c) monopole without DGS (Figure 9.25c); (d) monopole with DGS (Figure 9.25d). Source: Malik et al. [45] IEEE.

this is prominently visible in Figure 9.26d. This actually signifies an important factor to tackle any inter symbol interference in high-speed data communications.

9.6 Conclusion

The state-of-the-art developments and applications indicate that DGS has been an integrated part of UWB-printed monopole designs. Apart from providing safeguards as band notches in its impedance characteristics, it addresses the main challenge is optimizing its shape and deployment location in view of limited ground plane size. The DGS, in many cases, control the radiation patterns where small ground planes take prominent role in radiation. Hence, such designs require strong understandings on DGS fundamentals and a thorough knowledge on its working principle.

References

1 First Report and Order, "Revision of Part 15 of the commission's rule regarding ultra-wideband transmission system FCC02-48," Federal Communications Commission, 2002.

2 D. Guha and Y. Antar, Eds., *"Chapter 10 in Microstrip and Printed Antennas: New Trends, Techniques and Applications,"* Wiley, UK, 2011.

3 W. S. Chen and Y. H. Yu, "A CPW-fed rhombic antenna with band reject characteristics for UWB applications," *Microwave Journal*, vol. 50, no. 11, pp. 102–110, 2007.

4 J. Jung, W. Choi, and J. Choi, "A small wideband microstrip-fed monopole antenna," *IEEE Microwave and Wireless Components Letters*, vol. 15, no. 10, pp. 703–705, 2005.

5 R. Gayathri, T. U. Jisney, D. D. Krishna, M. Gopikrishna, and C. K. Aanandan, "Band-notched inverted-cone monopole antenna for compact UWB systems," *Electronic Letters*, vol. 44, no. 20, pp. 2008.

6 J. B. Jiang, Y. Song, Z. H. Yan, X. Zhang, and W. Wu, "Band-notched UWB printed antenna with an inverted-L-slotted ground," *Microwave and Optical Technology Letters*, vol. 51, no. 1, pp. 260–263, 2009.

7 J. B. Jiang, Z.-H. Yan, and J.-Y. Zhang, "Dual band-notched ultra-wideband printed antenna with two different types of slots," *Microwave and Optical Technology Letters*, vol. 52, no. 9, pp. 1930–1933, 2010.

8 L. Li, Z.-L. Zhou, J.-S. Hong, and B.-Z. Wang, "Compact dual-band-notched UWB planar monopole antenna with modified SRR," *Electronic Letters*, vol. 47, no. 17, 950–951, 2011.

9 D. Yadav, M. P. Abegaonkar, S. K. Koul, V. Tiwari, and D. Bhatnagar, "A novel frequency reconfigurable monopole antenna with switchable characteristics between band-notched UWB and WLAN applications," *Progress In Electromagnetics Research C*, vol. 77, pp. 145–153, 2017.

10 Q. Wu, R. Jin, J. Geng, and J. Lao, "Ultra-wideband rectangular disk monopole antenna with notched ground," *Electronic Letters*, vol. 43, no. 11, 605–606, 2007.

11 M. Jalali and T. Sedghi, "Very compact UWB CPW-fed fractal antenna using modified ground plane and unit cells," *Microwave and Optical Technology Letters*, vol. 56, no. 4, pp. 851–854, 2014.

12 F. Fereidoony, S. Chamaani, and S. A. Mirtaheri, "Systematic design of UWB monopole antennas with stable omnidirectional radiation pattern," *IEEE Antennas and Wireless Propagation Letters*, vol. 11, pp. 752–755, 2012.

13 K. H. Chiang and K. W. Tam "Microstrip monopole antenna with enhanced bandwidth using defected ground structure," *IEEE Antennas and Wireless Propagation Letters*, vol. 7, pp. 532–535, 2008.

14 C. C. Leong, W. W. Choi, and K. W. Tam, "A tunable monopole antenna using double U-shaped defected ground structure with islands," *International Symposium on Antennas and Propagation (ISAP)*, October, The Institute of Electronics, Information and Communication Engineers (IEICE), Thailand, 783–786, 2009.

15 M. A. Antoniades and G. V. Eleftheriades, "A compact monopole antenna with a defected ground plane for multiband applications," *2008 IEEE Antennas and Propagation Society International Symposium*, San Diego, CA, pp. 1–4, 2008.

16 M. A. Antoniades and G. V. Eleftheriades, "A compact multiband monopole antenna with a defected ground plane," *IEEE Antennas and Wireless Propagation Letters*, vol. 7, pp. 652–655, 2008.

17 J. Zhu, M. A. Antoniades, and G. V. Eleftheriades, "A compact tri-band monopole antenna with single-cell metamaterial loading," *IEEE Antennas and Wireless Propagation Letters*, vol. 58, no. 4, pp. 1031–1038, 2010.

18 M. A. Antoniades and G. V. Eleftheriades, "A broadband dual-mode monopole antenna using NRI-TL metamaterial loading," *IEEE Antennas and Wireless Propagation Letters*, vol. 8, pp. 258–261, 2009.

19 M. A. Antoniades and G. V. Eleftheriades, "A folded-monopole model for electrically small NRI-TL metamaterial antennas," *IEEE Antennas and Wireless Propagation Letters*, vol. 7, pp. 425–428, 2008.

20 Z. Y. Liu, Y. Z. Yin, S. F. Zheng, W. Hu, and L. H. Wen, "A compact CPW-fed monopole antenna with a U-shaped strip and a pair of L-slits ground for WLAN and WIMAX applications," *Progress in Electromagnetics Research Letters*, vol. 16, pp. 11–19, 2010.

21 W. C. Liu, C. M. Wu, and Y. Dai, "Design of triple-frequency microstrip-fed monopole antenna using defected ground structure," *IEEE Transactions on Antennas and Propagation*, vol. 59, no. 7, pp. 2457–2463, 2011.

22 M. Ojaroudi, C. Ghobadi, and J. Nourinia, "Small square monopole antenna with inverted T-shaped notch in the ground plane for UWB application," *IEEE Antennas and Wireless Propagation Letters*, vol. 8, pp. 728–731, 2009.

23 M. Rostamzadeh, S. Mohamadi, J. Nourinia, C. Ghobadi, and M. Ojaroudi, "Square monopole antenna for UWB applications with novel rod-shaped parasitic structures and novel V-shaped slots in the ground plane," *IEEE Antennas and Wireless Propagation Letters*, vol. 11, pp. 446–449, 2012.

24 W. T. Li, X. W. Shi, and Y. Q. Hei, "Novel planar UWB monopole antenna with triple band-notched characteristics," *IEEE Antennas and Wireless Propagation Letters*, vol. 8, pp. 1094–1098, 2009.

25 M. Abdollahvand, G. Dadashzadeh, and D. Mostafa, "Compact dual band-notched printed monopole antenna for UWB application," *IEEE Antennas and Wireless Propagation Letters*, vol. 9, pp. 1148–1151, 2010.

26 L.-X. Li, S.-S. Zhong, and M.-H. Chen, "Compact band-notched ultrawideband antenna using defected ground structure," *Microwave and Optical Technology Letters*, vol. 52, no. 2, pp. 286–289, 2010.

27 A. Nouri and G. R. Dadashzadeh, "A compact UWB band-notched printed monopole antenna with defected ground structure," *IEEE Antennas and Wireless Propagation Letters*, vol. 10, pp. 1178–1181, 2011.

28 S. Soltani, M. Azarmanesh, P. Lotfi, and G. Dadashzadeh, "Two novel very small monopole antennas having frequency band notch function using DGS for UWB application," *International Journal of Electronics and Communications*, vol. 65, no. 1, pp. 87–94, 2011.

29 J.-K. Lee and Y.-S. Kim, "A multiband-rejected UWB monopole antenna using interdigital defected ground structure," *Microwave and Optical Technology Letters*, vol. 53, no. 2, pp. 312–314, 2011.

30 J. J. Liu, K. P. Esselle, S. G. Hay, and S. S. Zhong, "Planar ultra-wideband antenna with five notched stop bands," *Electronics Letters*, vol. 49, no. 9, pp. 579–580, 2013.

31 Y. S. Li, X. D. Yang, C. Y. Liu, and T. Jiang, "Compact CPW-fed ultra-wideband antenna with dual band-notched characteristics," *IEICE Electronics Express*, vol. 7, no. 20, pp. 1597–1601, 2010.

32 J.-Y. Kim, B.-C. Oh, N. Kim, and S. Lee, "Triple band-notched UWB antenna based on complementary meander line SRR," *Electronics Letters*, vol. 48, no. 15, pp. 896–897, 2012.

33 M. Elhabchi, M. N. Srifi, and R. Touahni, "A tri-band-notched UWB planar monopole antenna using DGS and semi arc-shaped slot for WiMAX/WLAN/X-band rejection," *Progress In Electromagnetics Research Letters*, vol. 70, pp. 7–14, 2017.

34 C. Zhang, J. Zhang, and L. Li, "Triple band-notched UWB antenna based on SIR-DGS and fork-shaped stubs," *Electronics Letters*, vol. 50, no. 2, pp. 67–69, 2014.

35 R. Zaker, C. Ghobadi, and J. Nourinia, "Novel modified UWB planar monopole antenna with variable frequency band-notch function," *IEEE Antennas and Wireless Propagation Letters*, vol. 7, pp. 112–114, 2008.

36 Y. D. Dong, W. Hong, Z. Q. Kuai, and J. X. Chen, "Analysis of planar ultra-wideband antennas with on-ground slot band-notched structures," *IEEE Transactions on Antennas and Propagation*, vol. 57, no. 7, pp. 1886–1893, 2009.

37 Y. S. Li, X. D. Yang, C. Y. Liu, and T. Jiang, "Compact CPW-fed ultra-wideband antenna with band-notched characteristic," *Electronic Letters*, vol. 46, no. 23, pp. 1533–1534, 2010.

38 S. C. Puri, S. Das, and M. G. Tiary, "An UWB trapezoidal rings fractal monopole antenna with dual-notch characteristics," *Microwave and Optical Technology Letters*, vol. 29, no. 8, 2019.

39 Z.-P. Zhong, J. J. Liang, M. L. Fan, G. L. Huang, W. He, X. C. Chen, and T. Yuan, "A compact CPW-fed UWB antenna with quadruple rejected band," *Microwave and Optical Technology Letters*, vol. 61, no. 12, pp. 2795–2800, 2019.

40 S. C. Puri, S. Das, and M. G. Tiary, "UWB monopole antenna with dual-band-notched characteristics," *Microwave and Optical Technology Letters*, vol. 62, no. 3, pp. 1222–1229, 2020.

41 M. C. Derbal, A. Zeghdoud, and M. Nedil, "A dual band notched UWB antenna with optimized DGS using genetic algorithm," *Progress In Electromagnetics Research Letters*, vol. 88, pp. 89–95, 2020.

42 J. Zhu, B. Feng, B. Peng, L. Deng, and S. Li, "A dual notched band MIMO slot antenna system with Y-shaped defected ground structure for UWB application," *Microwave and Optical Technology Letters*, vol. 58, no. 3, pp. 626–630, 2016.

43 G. Liu, Y. Liu, and S. Gong, "Compact uniplanar UWB MIMO antenna with band-notched characteristic," *Microwave and Optical Technology Letters*, vol. 59, no. 9, pp. 2207–2212, 2017.

44 Z. Tang, X. Wu, J. Zhan, S. Hu, Z. Xi, and Y. Liu, "Compact UWB-MIMO antenna with high isolation and triple band-notched characteristics," *IEEE Access*, vol. 7, pp. 19856–19865, 2019.

45 J. Malik, A. Patnaik, and M. V. Kartikeyan, "Time-domain performance of band-notch techniques in UWB antenna," in *2016 Asia-Pacific Microwave Conference (APMC)*, New Delhi, India, pp. 1–3, 2016.

Index

Printed and bound by CPI Group (UK) Ltd, Croydon, CR0 4YY

16/04/2025